THE NAMES OF PLANTS

THE NAMES OF PLANTS

D. GLEDHILL
Department of Botany, University of Bristol

The right of the
University of Cambridge
to print and sell
all manner of books
was granted by
Henry VIII in 1534.
The University has printed
and published continuously
since 1584.

CAMBRIDGE UNIVERSITY PRESS
Cambridge
London   New York   New Rochelle
Melbourne   Sydney

Published by the Press Syndicate of the University of Cambridge
The Pitt Building, Trumpington Street, Cambridge CB2 1RP
32 East 57th Street, New York, NY 10022, USA
10 Stamford Road, Oakleigh, Melbourne 3166, Australia

First published 1985

Printed in Great Britain by
the University Press, Cambridge

Library of Congress catalogue card number: 85-47881

British Library cataloguing in publication data
Gledhill, D.
The names of plants.

1. Botany – Nomenclature
I. Title
581'.014   QK96

ISBN 0 521 30549 7 hard covers
ISBN 0 521 31562 X paperback

Contents

Preface

Originally entitled *The naming of plants and the meanings of plant names*, this book is in two parts. The first part has been written as an account of the way in which the naming of plants has changed with time and why the changes were necessary. It has not been my intention to dwell upon the more fascinating aspects of common names but rather to progress from these to the situation which exists today, in which all plant names must conform to internationally agreed standards. I have aimed at producing an interesting text which is equally as acceptable to the amateur gardener as to the botanist.

The book had its origins in a collection of Latin plant names and their meanings in English, which continued to grow by the year but which could never be complete. The disparities which I noted, between those names which had meaningful translations and those which did not, were added to by the disparities found in some books which not only gave full citation of plant names but also gave a translation of the name, as well as a common name, and others which did neither, and by a tendency which may be part of modern language to reduce the names, especially of garden plants, to an abbreviated form (e.g. *Rhodo* for *Rhododendron*). Having myself produced such meaningless common names as Vogel's napoleona and Ivory Coast mahogany by translation of the botanical names (*Napoleona vogelii* and *Khaya ivorensis* respectively), I have presented a glossary which should serve to translate the more meaningful, descriptive names of plants from anywhere on earth but which will give little information about the many people and places epitomized in plant names.

I make no claim that the meanings which I have listed are always the only meanings which have been put upon the various entries. Authors of Latin names have not always explained the meanings of the names which they have erected and as a result such names have subsequently been given different meanings.

The nature of the problem

A rose? By any name?

Man's highly developed constructive curiosity and his capacity for communication are two of the attributes which distinguish him from all other animals. Man alone has sought to understand the whole living world, and things beyond his own environment, and to pass his knowledge on to others. Consequently, when he discovers or invents something new he also creates a new word, or words, in order to be able to communicate his discovery or invention to others. There are no rules to govern the manner in which such new words are formed other than those of their acceptance and acceptability. This is equally true of the common, or vulgar, or vernacular names of plants. Such names present few problems until communication becomes multilingual and the number of plants named becomes excessive. For example, the diuretic dandelion is easily accommodated in European languages as the lion's tooth (Löwenzahn, dent de lion, dente di leone) or as piss-abed (Pissenlit, piscacane, piscialetto) but when further study reveals that there are more than a thousand different kinds of dandelion throughout Europe then formulation of common names for these is both difficult and unacceptable.

Common plant names present language at its richest and most imaginative (welcome home husband however drunk you be, for the houseleek or *Sempervivum*; shepherd's weather-glass, for scarlet pimpernel or *Anagallis*; meet her i'th'entry kiss her i'th'buttery, or leap up and kiss me, for *Viola tricolor*; touch me not, for the balsam *Impatiens noli-tangere*: mind your own business, or mother of

1

thousands, for *Soleirolia soleirolii;* blood drop emlets, for *Mimulus luteus*). Local variations in common names are numerous and this is perhaps a reflection of the importance of plants in general conversation, in the kitchen and in herbalism throughout the country in bygone days. An often-quoted example of the multiplicity of vernacular names is that of *Caltha palustris* for which, in addition to marsh marigold, kingcup, and May blobs, there are another 90 local British names (one being dandelion); as well as over 140 German vernacular names and 60 French ones.

Common plant names have many sources. Some came from antiquity by word of mouth as part of language itself and the passage of time and changing circumstances have obscured their meanings. Fanciful ideas of a plant's association with animals, ailments, and festivities, and observation of plant structures, perfumes, colours, habitats and seasonality have all contributed to their naming. So also have their names in other languages. English plant names have come from Arabic, Persian, Greek, Latin, ancient British, Anglo-Saxon, Norman, Low German, Swedish and Danish. Such names were introduced together with the spices, grains, fruit plants and others which merchants and warring nations introduced to new areas. Foreign names often remained little altered but some were transliterated in such a way as to lose any meaning which they may have had originally.

The element of fanciful association in vernacular plant names often drew upon comparisons with parts of the body and with bodily functions (priest's pintle for *Arum maculatum*, open arse for *Mespilus germanicus* and arse smart for *Polygonum hydropiper*). Some of these persist but no longer strike us as 'vulgar' because they are 'respectably' modified or the associations themselves are no longer familiar to us (*Arum maculatum* is still known as cuckoo pint (cuckoo pintle) and as wake robin). Such was the sensitivity to indelicate names that Britten and Holland, in their *Dictionary of English Plant Names* (1886), wrote 'We have also purposely

2

excluded a few names which, though graphic in their construction and meaning, interesting in their antiquity, and even yet in use in certain counties, are scarcely suited for publication in a work intended for general readers'. They nevertheless included the examples above. The cleaning up of such names was a feature of the Victorian period, during which our common plant names were formalized and reduced in numbers. Some of the resulting names are prissy (bloody cranesbill becomes blood-red cranesbill), some are uninspired (naked ladies or meadow saffron or *Colchicum autumnale* becomes Autumn Crocus) and most are not very informative.

This last point is not of any real importance because names do not need to have a meaning or be interpretable. Primarily, names are mere cyphers which are easier to use than lengthy descriptions and yet, when accepted, they can be quite as meaningful. Within limits, it is possible to use one name for a number of different things but, if the limits are exceeded, this may cause great confusion. There are many common plant names which refer to several plants but cause no problem so long as they are used only within their local areas or when they are used to convey only a general idea of the plant's identity. For example, *Wahlenbergia saxicola* in New Zealand, *Phacelia whitlavia* in southern California, U.S.A., *Clitoria ternatea* in West Africa, *Campanula rotundifolia* in Scotland, and *Endymion non-scriptus* (formerly *Scilla non-scripta* and now *Hyacinthoides non-scripta*) in England are all commonly called bluebells. In each area, local people will understand others who speak of bluebells but in all the areas except Scotland the song *The Bluebells of Scotland*, heard perhaps on the radio, will conjure up a wrong impression. At least ten different plants are given the common name of cuckoo-flower in England, signifying only that they flower in spring at a time when the cuckoo is first heard.

The problem of plant names and plant naming is that common names need not be formed according to any rule and can change as language, or the user of language, dictates. If our awareness extended only to some thousands of 'kinds' of plants we could manage by giving them numbers but, as our awareness extends, more 'kinds' are recognized and we find a need to organize our thoughts about them by giving them names and by forming them into named groups. Then we have to agree with others about the names and the groups, otherwise communication becomes hampered by ambiguity.

Formalizing names provides a partial solution to the two opposed problems presented by vernacular names, multiple naming of a single plant and multiple application of a single name. The predominantly two-word structure of such formal names has been adopted in recent historic times in all biological nomenclature especially in that branch which, thanks to Isidorus Hispalensis of Seville (560–636), we now call botany. Of necessity, botanical names have been formulated from former common names but this does not mean that in the translation of botanical names we may expect to find meaningful names in common language. Botanical names, however, do represent a stable system of nomenclature which is usable by people of all nationalities and has relevancy to a system of classification.

The size of the problem

Three centuries before Christ, Aristotle, disciple of Plato, wrote extensively and systematically of all that was then known of the physical and living world. In this monumental task he laid the foundations of inductive reasoning. When he died, he left his writings and his teaching garden to one of his pupils, Theophrastus (380–287 B.C.), who also took over Aristotle's peripatetic school. Theophrastus' writings on mineralogy and plants totalled 227 treatises, of which nine books of *Historia Plantarum* and eight of *De Causis Plantarum* were subsequently translated into Syrian, to Arabic, to Latin and back to Greek. His writings were based upon his own critical observations, a departure from earlier philosophical approaches, and rightly entitle him to be regarded as the father of botany. He recognized the distinctions between monocotyledons and dicotyledons, superior and inferior ovaries in flowers, the necessity for pollination and the sexuality of plants.

To the ancients, as to the people of earlier civilizations of Persia and China, plants were distinguished on the basis of their culinary, medicinal and decorative uses – as well as their supposed supernatural properties. For this reason plants were given a name and also a description. Theophrastus wrote of some 500 'kinds' of plant which, considering that he had material brought back from Alexander the Great's campaigns, would indicate a considerable lack of discrimination. In Britain we now recognize more than that number of different 'kinds' of moss.

Four centuries later, about A.D. 64, Dioscorides recorded

600 'kinds' of plants and, half a century later still, the elder Pliny described about 1000 'kinds'. During the 'Dark Ages', despite the remarkable achievements of such people as Albertus Magnus (1193–1280) who collected plants during extensive journeys in Europe, and the publication of the *German Herbarius* in 1485 by another collector of European plants, Dr Johann von Cube, little progress was made in the study of plants. It was the renewal of critical observation by Renaissance botanists such as Dodoens (1517–1585), l'Obel (1538–1616), l'Ecluse (1526–1609) and others which resulted in the recognition of some 4000 'kinds' of plants by the sixteenth century. At this point in history, the renewal of critical study and the beginning of plant collection throughout the known world produced a requirement for a rational system of grouping plants. Up to the sixteenth century three factors had hindered such classification. The first of these was that the main interested parties were the nobility and apothecaries who conferred on plants great monetary value either because of their rarity or because of the real or imaginary virtues attributed to them and regarded them as items to be guarded jealously. Second was the lack of any standardized system of naming plants, and third, and perhaps most important, any expression of the idea that living things could have evolved from earlier extinct ancestors and could therefore form groupings of related 'kinds' was a direct contradiction of the religious dogma of Divine Creation.

Perhaps the greatest disservice to progress was that caused by the doctrine of signatures, which claimed that God had given to each 'kind' of plant some feature which would indicate the uses to which man could put the plant. Thus, plants with kidney-shaped leaves could be used for treating kidney complaints and were grouped together on this basis. Theophrastus Bombast von Hohenheim (1493–1541) had invented properties for many plants under this doctrine and also considered that man had intuitive knowledge of which

6

plants could serve him, and how. He is better known under the Latin name which he assumed, Paracelsus, and the doctrinal book *Dispensatory* is usually attributed to him. The doctrine was also supported by Giambattista Della Porta (1543–1615) who made an interesting extension to it: that the distribution of different 'kinds' of plants had a direct bearing upon the distribution of different kinds of ailment which man suffered in different areas. On this basis, the preference of willows for wet habitats is ordained by God because men who live in wet areas are prone to suffer from rheumatism and, since the bark of *Salix* species gives relief from rheumatic pains (it contains salicylic acid, the analgesic principal of aspirin) the willows are there to serve the needs of man.

In spite of disadvantageous attitudes, renewed critical interest in plants during the sixteenth century led to more discriminating views as to the nature of 'kinds', to searches for new plants from different areas and concern over the problems of naming plants. John Parkinson (1569–1629) a London apothecary, wrote a horticultural landmark in his *Paradisi in Sole* of 1629. This was an encyclopaedia of gardening and of plants then in cultivation and contains a lament by Parkinson that, in their many catalogues, nurserymen 'without consideration of kinde or form, or other special note, give(th) names so diversely one from the other, that...very few can tell what they mean'. This attitude towards common names is still with us but not in so violent a guise as that shown by an unknown writer who, in *Science Gossip* of 1868, wrote that vulgar names of plants presented 'a complete language of meaningless nonsense, almost impossible to retain and certainly worse than useless when remembered – a vast vocabulary of names, many of which signify that which is false, and most of which mean nothing at all'.

Names continued to be formed as phrase-names with a single word name (which was to become the generic name) followed by a description. So, we find that the creeping buttercup was known by many names, of which Caspar Bauhin (1550–1664) and Christian Mentzel (1622–1701) listed the following:

Caspar Bauhin, *Pinax Theatri Botanici*, 1623

Ranunculus pratensis repens hirsutus var. C. Bauhin
   repens fl. luteo simpl. J. Bauhin
   repens, fol. ex albo variis
   repens magnus hirsutus fl. pleno
   repens flore pleno
   pratensis repens Parkinson
   pratensis reptante cauliculo l'Obel
   polyanthemos 1 Dodoens
   hortensis 1 Dodoens
   vinealis Tabernamontana
   pratensis etiamque hortensis Gerard

Christianus Mentzelius, *Index Nominum Plantarum Multilinguis (Universalis)*, 1682

Ranunculus pratensis et arvensis C. Bauhin
   rectus acris var. C. Bauhin
   rectus fl. simpl. luteo J. Bauhin
   rectus fol. pallidioribus hirsutis J. Bauhin
   albus fl. simpl. et denso J. Bauhin
   pratensis erectus dulcis C. Bauhin
Ranoncole dolce Italian
Grenoillette dorée ò doux Gallic
Sevvite Woode Cravve foet English
Suss Hanènfuss
Jaskien sodky Polish
Chrysanth. simplex Fuchs
Ranunculus pratensis repens hirsutus var. C C. Bauhin
   repens fl. luteo simpl. J. Bauhin
   repens fol. ex albo variis Antonius Vallot
   repens magnus hirsut. fl. pleno J. B. Tabernamontana
   repens fl. pleno J. Bauhin
   arvensis echinatus Paulus Ammannus
   prat. rad. verticilli modo rotunda. C. Bauhin
   tuberosus major J. Bauhin

Crus Galli Otto Brunfelsius
Coronopus parvus, Batrachion Apuleius Dodonaeus (Dodoens)
Ranunculus prat. parvus fol. trifido C. Bauhin
      arvensis annuus fl. minimo luteo Morison
      fasciatus Henricus Volgnadius
      Ol. Borrich Caspar Bartholino

These were, of course, common or vernacular names with wide currency and strong candidates for inclusion in lists which were intended to clarify the complicated state of plant naming. Local, vulgar names escaped such listing until much later times, when they were being less used and lexicographers began to collect them, saving most from vanishing for ever.

During the seventeenth century great advances were made. Robert Morison (1620–1682) published a convenient, or artificial, system of grouping 'kinds' into groups of increasing size, as a hierarchy. One of his groups we now call the family Umbelliferae or, to give it its modern name, Apiaceae, and this was the first natural group to be recognized. By natural group we imply that the members of the group share a sufficient number of common features to suggest that they have all evolved from a single ancestral stock. Joseph Pitton de Tournefort (1656–1708) had made a very methodical survey of plants and had assorted 10000 'kinds' into 698 groups (or genera). The 'kinds' must now be regarded as the basic units or classification called *species*. Although critical observation of structural and anatomical features led to classification advancing beyond the vague herbal and signature systems no such advance was made in plant naming until a Swede, of little academic ability when young, we are told, established landmarks in both classification and nomenclature of plants. He was Carl Linnaeus (1707–1778) who classified 7300 species in 1098 genera and gave to each species a binomial name (a name consisting of a generic name-word plus a descriptive epithet, both of Latin form).

It was inevitable that, as man grouped the ever-increasing number of known plants (and he was yet only aware of European, Mediterranean and a few plants from other areas) the constancy of associated morphological features in some groups should suggest that the whole was derived, by evolution, from a common ancestor. Morison's family Umbelliferae was a case in point. Also, because the basic unit of any system of classification is the species, and some species were found to be far less constant than others, it was just as inevitable that the nature of the species itself would become a matter of controversy, not least in terms of religious dogma. A point often passed over with insufficient comment is that Linnaeus' endeavours towards a natural system of classification were accompanied by a changing attitude towards Divine Creation. His early view was that true species were produced by the hand of the Almighty and that abnormal varieties were produced by nature in a sporty mood. In such genera as *Thalictrum* and *Clematis* he concluded that some species were not original creations and, in *Rosa*, he was drawn to conclude that some species had either blended or that one species had given rise to several others. Later, he invoked hybridization as the process by which species could be created and attributed to the Almighty the creation of the primeval genera, each with a single species. From the maxims by which he expressed his views in *Genera Plantarum* 6th edn (1764) he attributed the creation of a basic plant, the blending from it of the orders (our families) and from these of the genera to the Creator. Species, he went on, were produced by nature, by hybridization between the genera. The abnormal varieties of the species so formed were the product of chance.

Linnaeus was well aware of the kind of results which plant hybridizers were obtaining in Holland and it is not surprising that his own knowledge of naturally occurring variants led him towards a covertly expressed belief in evolution. However, that expression, and his listing of varieties under

10

their typical species in *Species Plantarum*, where he indicated each with a Greek letter, was still contrary to the dogma of Divine Creation and it would be another century before an authoritative declaration of evolutionary theory was to be made, by Charles Darwin.

Darwin's essay on *The Origin of Species by Means of Natural Selection* (1859) was published somewhat reluctantly and in the face of fierce opposition. It was concerned with the major evolutionary changes by which species evolve and was based upon Darwin's own observations on fossils and living creatures. The concept of natural selection, or the survival of any life-form being dependent upon its ability to compete successfully for a place in nature, became, and still is, accepted as the major force directing an inevitable process of organic change. Our conception of the mechanisms and causative factors for the larger evolutionary steps, such as the demise of the dinosaurs and the many plant groups, now known only as fossils, and the emergence and diversification of the flowering plants during the last 100 million years, is, at best, hazy.

The great age of plant-hunting, from the second half of the eighteenth century through most of the nineteenth century, produced a flood of species not previously known. Strange and exotic plants were once prized above gold and caused theft, bribery and murder. Trading in 'paper tulips' by the van Bourse family gave rise to the continental stock exchange – the Bourse. With the invention of the Wardian Case by Dr Nathaniel Bagshot Ward, in 1827, it became possible to transport plants from the farthest corners of the world by sea and without enormous losses. The case was a small glasshouse which reduced water losses, and made it unnecessary to use large quantities of fresh water on the plants during long sea voyages, and also gave protection from salt spray. In the confusion which was created in naming this flood of plants, and in using many languages to describe them, it became apparent that there was a need for international agreement on both these matters. Today, the rules which have been formulated govern the names of about 300000 species of plants, which are now generally accepted, and have disposed of a great number of names which were found not to be valid.

Our present state of knowledge about the mechanisms of inheritance and change in plants and animals is almost entirely limited to an understanding of the cause of variation within a species. That understanding is based upon the observed behaviour of inherited characters as first recorded in *Pisum* by Gregor Johann Mendel, in 1866. With the technical development of the microscope, Malpighi (1641), Grew (1672) and others had explored the cellular structure of plants and elucidated the mechanism of fertilization. However, the nature of inheritance and biological variability remained clouded by myth and monsters until Mendel's work was rediscovered at the beginning of the present century. By 1900, de Vries, Correns, Tschermak and Bateson had confirmed that inheritance had a definite, particulate character which is regulated by 'genes'. Sutton (1902) was the first person to clarify the manner in which the characters are transmitted from parents to offspring when he described the behaviour of the 'chromosomes' during division of the cell nucleus. Chromosomes are thread-like bodies which can be stained and observed during division of the cell nucleus. Along their length, it can be shown, the sites of genetic control, or genes, are situated in an ordered linear sequence. Differences between individuals can now be explained in terms of the different forms, or allelomorphs, in which single genes can exist as a consequence of their mutation.

The concept of a taxonomic species, or grouping of individuals each of which has a close resemblance to the others in every aspect of its morphology, and to which a name can be applied, is not always the most accurate interpretation of the true circumstances in nature. It defines and delimits an entity but we are constantly discovering that the species is far from being an immutable entity. The botanist discovers that a species has components which have well-defined individual properties (an ability to live on a distinctive soil type, or an adaptation to flower and fruit in harmony with some agricultural practice, or having reproductive barriers caused by differences in chromosome number, etc.) and the horticultural plant breeder produces a steady stream of new varieties of cultivated species.

If we consider some of the implications of, and attitudes towards, delimiting plant species and their components, and naming them, it will become easier to understand the need for the internationally accepted rules intended to prevent the unnecessary and unacceptable proliferation of names.

Towards a solution to the problem

It is basic to the collector's art to arrange items into groups. Postage stamps can be arranged by country of origin and then on face value, year of issue, design, colour variations or defects. The arranging process always resolves into a hierarchic set of groups. In the plant kingdom we have a descending hierarchy of groups through divisions, divided into classes, divided into orders, divided into families, divided into genera, divided into species. Subsidiary groupings are possible at each level of this hierarchy and are employed to rationalize the uniformity of relationships within the particular group. Thus, a genus may be divided into a mini-hierarchy of sub-genera, divided into sections, divided into series in order to assort the components into groupings of close relatives. All the components would, nevertheless, be members of the one genus. Early systems of classification were much less sophisticated and were based upon few aspects of plant structure, such as those which suggested signatures, and mainly upon ancient herbal–medicinal concepts.

Otto Brunfels (1488–1534) was probably the first person to introduce accurate, objective recording and illustration of plant structure in his *Herbarium* of 1530–36, and Valerius Cordus (1515–1544) could have revolutionized botany but for his premature death. His four books of German plants contained detailed accounts of the structure of 446 plants, based on his own systematic studies on them. Many of the plants were new to science. A fifth book on Italian plants was in compilation when he died. Conrad Gesner (1516–1565) published Cordus' works on German plants in 1561 and the fifth book in 1563.

A primitive suggestion of an evolutionary sequence was contained in Matthias de l'Obel's *Plantarum seu Stirpium Historia* (1576) in which narrow-leaved plants, followed by broader-leaved bulbous and rhizomatous plants, followed by herbaceous dicotyledons, followed by shrubs and trees, was regarded as a series of increasing 'perfection'. Andrea Caesalpino (1519–1603) retained the distinction between woody and herbaceous but employed more detail of flower, fruit and seed structure in compiling his classes of plants (*De Plantis*, 1583). His influence extended to the classifications of Caspar Bauhin (1550–1624), who departed from the use of medical information and compiled detailed descriptions of the plants to which he gave many binomial names; P. R. de Belleval (1558–1632), who had adopted a naming system consisting of Latin nouns plus Greek adjectival words as binomials; Joachim Jung (1587–1657), whose fear of accusation of heresy prevented him from publishing his work, but some of whose manuscripts survived – these contain many of the terms which we still use in describing leaf and flower structure and arrangement, and also contain plant names consisting of a noun qualified by an adjective; Robert Morison (1620–1683), mentioned earlier; John Ray (1686–1704), who introduced the distinction between mono-cotyledons and dicotyledons but retained the distinction between flowering herbaceous plants and woody plants, and also used binomial names; and Carl Linnaeus who deserves separate comment.

Joseph Pitton de Tournefort (1656–1708) placed great emphasis on the floral corolla and upon defining the genus, rather than the species. His 698 generic descriptions are detailed but his species descriptions are dependent on binomials and illustrations. Herman Boerhaave (1668–1739) combined the systems of Ray and Tournefort, and others, to incorporate morphological, ecological, leaf and floral/fruiting characters, but none of these early advances received popular support. As Michel Adanson (1727–1806) was to realize, some sixty systems of classification had been proposed by the middle of the eighteenth century and none had been free from narrow conceptual restraints. His plea that attention should be focused on 'natural' classification through processes of inductive reasoning, because of the wide range of characteristics then being employed, did not enjoy wide publication and his work was not well regarded when it did become more widely known.

Before considering the major contributions made by Carl Linnaeus, it should be noted that the names of many higher groups of plants, of families and genera were well established at the beginning of the eighteenth century and several people had used simplified, binomial names for species. Indeed, August Quirinus Rivinus (1652–1723) had proposed that no plant should have a name of more than two words.

Carl Linnaeus (1707–1778) was the son of a clergyman, Nils, who had adopted the latinized family name when he became a student of theology. Carl also went to theological college for a year but then left and became an assistant gardener in Prof. Olaf Rudbeck's botanic garden at Uppsala. His ability as a collector and arranger soon became evident and, after undertaking tours through Lapland and Holland, he began to publish works which are now the starting points for naming plants and animals. In literature he is referred to as Carl or Karl or Carolus Linnaeus, Carl Linné (an abbreviation) and, later in life, as Carl von Linné. His life became one of devotion to the classification and naming of all living things and of teaching others about them. His numerous students played a very important part in the discovery of new plants from many parts of the world.

Linnaeus' main contribution to botany was his method of naming plants, in which he combined Bauhin's and Belleval's use of binomials with Tournefort's and Boerhaave's concepts of the genus. His success, where all others before him had

16

failed, was due to his early publication of his most popular work, an artificial system of classifying plants, in which he employed the number and disposition of the stamens and the composition of the pistil of the flower to define 24 classes. This sexual system provided an easy way of grouping plants and of allocating newly discovered plants to a group. Originally designed to accommodate the plants of his home parish, it was elaborated to include first the Arctic flora and later the more diverse and exotic plants being discovered in the tropics. It continued in popular use into the nineteenth century despite its limitation of grouping together strange bedfellows; red valerian, tamarind, crocus, iris, galingale sedge and mat grass are grouped together under *Triandria* (three stamens) Monogynia (pistil with a single style).

In 1735, Linnaeus published *Systema Naturalis*, in which he grouped species into genera, genera into orders and orders into classes on the basis of structural similarities. This was an attempt to interpret evolutionary relationships or assemblages of individuals at various levels. In his *Species Plantarum*, published in 1753, he gave each species a binomial name. The first word of each binomial was the name of the genus to which the species belonged and the second word was a descriptive epithet, or specific epithet. Thus, the creeping buttercup he names as *Ranunculus repens.*

It now required that the systematic classification and the binomial nomenclature which Linnaeus had started should become generally accepted and, largely because of the popularity of his sexual system, this was to be the case. Botany could now contend with the rapidly increasing number of species of plants being collected for scientific inquiry, rather than for medicine or exotic gardening as in the seventeenth century. For the proper working of such standardized nomenclature, however, it was necessary that the language of plant names should also be standardized.

Linnaeus' views on the manner of forming plant names, and the use of Latin for these and for the description of plants and their parts, have given rise directly to modern practice and a Latin vocabulary of great versatility, but which would have been largely incomprehensible in ancient Rome. He applied the same methodical principles to the naming of animals, minerals and diseases and, in doing so, established Latin, the *lingua franca* of his day, as the internationally used language of those branches of science.

The rules by which we now name plants depend largely on Linnaeus' writings but for the names of plant families we are much dependent on A. L. de Jussieu's classification in his *Genera Plantarum* of 1789. For the name of species, the correct name is that which was first published since 1753. This establishes Linnaeus' *Species Plantarum* (associated with his *Genera Plantarum* 5th edn of 1754 and 6th edn of 1764) as the starting point for the names of species (and their descriptions).

Linnaeus' sexual system of classification was very artificial and, although Linnaeus must have been delighted at its popularity, he regarded it as no more than a convenient pigeon-holing system. He published some of his views on grouping plant genera into natural orders (our families) in *Philosophia Botanica* (1751). Most of his orders were not natural groupings but considerably mixed assemblages. By contrast, Bernard de Jussieu (1699–1777), followed by his nephew Antoine Laurent de Jussieu (1748–1836), searched for improved ways of arranging and grouping plants as natural groups. In A. L. de Jussieu's *Genera Plantarum* the characteristics are given for 100 plant families; and most of these we still recognize.

Augustin Pyrame de Candolle (1778–1841) also sought a natural system, as did his son Alphonse, and he took the evolutionist view that there is an underlying state of symmetry in the floral structure which we can observe today and that by considering relationships in terms of the symmetry, natural alliances may be recognized. This approach resulted in a great deal of monographic work from which de Candolle formed views on the concept of a core of similarity, or type, for any natural group and the requirement for control in the naming of plants.

Today, technological and scientific advances have made it possible for us to use sub-cellular, chemical and the minutest morphological features and to incorporate as many items of information as are available about a plant in computer-aided assessments of that plant's relationships to others. Biological information has often been found to conflict with the concept of the taxonomic species and there are many plant groups in which the 'species' can best be regarded as a collection of highly variable populations. The gleaning of new evidence necessitates a continuing process of reappraisal of families, genera and species. Such reappraisal may result in subdivision or even splitting of a group into several new ones or, the converse process, in lumping together two or more former groups into one new one. Since the bulk of research is carried out on individual species, most of the revisions are carried out at or below the rank of species. On occasion, therefore, a revision at the family level will require the transfer of whole genera from one family to another, but it is now more common for a revision at the level of the genus to require the transfer of some, if not all the species from one genus to another. Such revisions are not mischievous but are the necessary process by which newly acquired knowledge is incorporated into a generally accepted framework. It is because we continue to improve the extent of our knowledge of plants that revision of the systems for their classification continues and, consequently, that name changes are inevitable.

The equivalence, certainly in evolutionary terms, of groups of higher rank than the family is a matter of philosophical debate and, even at the family level we find divergence of views as to whether those with few components are equivalent to those with many components. Because the taxonomic species is the basic unit of any system of classification, we have to assume parity between species; that is to say, we assume that a widespread species is in every way comparable with a rare species which may be restricted in its distribution to a very small area.

It is a feature of plants that their diversity of habit, longevity, mode of reproduction and tolerance to environmental conditions presents a wide range of biologically different circumstances. For the taxonomic problem of delimiting, defining and naming a species we have to identify a grouping of individuals whose characteristics are sufficiently stable to be defined, in order that a name can be applied to the group and a 'type', or exemplar, can be specified for that name. It is because of this concept of the 'type' that changes have to be made in names of species in the light of new discoveries, and that entities below the rank of species have to be recognized. Thus, we speak of botanical 'subspecies' when part of the species grouping can be distinguished as having a number of features which remain constant and as having a distinctive geographical or ecological distribution. When the degree of departure from the typical material is of lesser order we may employ the inferior category of 'variety'. The term 'form' is employed to describe a variant which is distinct in a minor way only, such as a single feature difference which might appear sporadically due to genetic mutation or sporting.

When a new name has to be given to a plant which is widely known under its superseded old name, it is always a cause of some annoyance. Gardeners always complain about such name changes but there is no novelty in that. On the occasion of Linnaeus being proposed for Fellowship of the Royal Society, Peter Collinson wrote to him in praise of his *Species Plantarum* but, at the same time, complained that Linnaeus had introduced new names for so many well-known plants.

The gardener has some cause to be aggrieved by changes in botanical names. Few gardeners show much alacrity in adopting new names and perusal of gardening books and catalogues shows that horticulture seldom uses botanical names with all the exactitude which they can provide. Horticulture, however, not only agreed to observe the international rules of botanical nomenclature but also formulated its own additional rules for the naming of plants grown under cultivation. It might appear as though the botanist realizes that he is bound by the rules, whereas the horticulturalist does not, but to understand this we must recognize the different facets of horticulture. The rules are of greatest interest and importance to specialist plant breeders and gardeners with a particular interest in a certain plant group. For the domestic gardener it is the growing of beautiful plants which is the motive force behind his activity. Between the two extremes lies every shade of interest and the main emphasis on names is an emphasis on garden names. Roses, cabbages, carnations and leeks are perfectly adequate names for the majority of gardeners but if greater precision is needed, a gardener wishes to know the name of the variety. Consequently, most gardeners are satisfied with a naming system which has no recourse to the botanical rules whatsoever. Not surprisingly, therefore, seed and plant catalogues also avoid botanical names, and so do 'popular' gardening books. The specialist plant breeder, however, shows certain similarities to the apothecaries of an earlier age. Like them he guards his art and his plants jealously because they represent the source of his future income and, also like them, he has the desire to understand every aspect of his plants. The apothecaries gave us the first centres of botanical enquiry and the plant breeders of today give us the new varieties which are needed to satisfy our gardening and food-production requirements.

Botanical names occasionally have to be resorted to by gardeners when they discover some cultural problem with a plant which shares the same common name with several different plants. Lately, the Guernsey lily, around which has alway hung a cloud of mystery, has been offered to the public in the form of *Amaryllis belladonna* L. The true Guernsey lily has the name *Nerine sarniensis* Herb. (but was named *Amaryllis sarniensis* by Linnaeus). The epithet *sarniensis* means 'of Sarnia' or 'of Guernsey'; Sarnia was the old name for Guernsey; and is an example of a misapplied geographical epithet, since the plant's native area is South Africa. Some would regard the epithet as indicating the fact that Guernsey was the first place in which this lily was cultivated. This is historically incorrect, however, and does nothing to help the gardener who finds that the Guernsey lily which he has bought does not behave, in culture, as *Nerine sarniensis* is known to behave. This example is one involving a particularly contentious area as to the taxonomic problems of generic boundaries and typification but there are many others in which common and 'Latin' garden names are used for whole assortments of garden plants, ranging from species (*Nepeta mussini* and *N. cataria* are both catmint) to members of different genera (japonicas including *Chaenomeles speciosa* and *Kerria japonica*) to members of different families (*Camellia japonica* is also a japonica, like the last two species above, and the diversity of bluebells was mentioned earlier).

New varieties, be they timber trees, crop plants or garden flowers, require names and those names need to be definitive. As with the earlier confusion of botanical names (different names for the same species or the same name for different species), so there can be the same confusion of horticultural names. As will be seen, rules for cultivated plants require that new names have to be established by publication. This gives to the breeder the commercial advantage of being able to supply to the public his new variety under what, initially, amounts to his mark of copyright. Indeed, in some parts of the world the breeder's new varietal name actually represents a trade mark.

The rules of botanical nomenclature

The rules which now govern the naming and the names of plants really had their beginnings in the views which A. P. de Candolle expressed in his *Théorie Elémentaire de la Botanique* (1813). There, he advised that plants should have names in Latin (or Latin form but not compounded from different languages), formed according to the rules of Latin grammar and subject to the right of priority for the name given by the discoverer or the first describer. This advice was found to be inadequate and, in 1862, the International Botanical Congress in London adopted control over agreements on nomenclature. Alphonse de Candolle (1806–1893), who was A. P. de Candolle's son, drew up four simple 'Lois', or laws, which were aimed at resolving what threatened to become a chaotic state of plant nomenclature. The Paris International Botanical Congress of 1867 adopted the Lois, which were:

1. One plant species shall have no more than one name.
2. No two plant species shall share the same name.
3. If a plant has two names, the name which is valid shall be that which was the earliest one to be published after 1753.
4. The author's name shall be cited, after the name of the plant, in order to establish the sense in which the name is used and its priority over other names.

It can be seen from these first rules that the nineteenth-century botanist frequently gave names to plants with little regard to either the previous use of the same name or the existing names which had already been applied to the same

23

plant. It is because of this aspect that one often encounters the words 'sensu' and 'non' inserted before the name of an author, although both terms are more commonly used in the sense of taxonomic revision, and indicate that the name is being used 'in the sense of' (sensu) or 'not in the sense of' (non) that author.

The use of Latin, as the language in which descriptions and diagnoses were written, was not universal in the nineteenth century and many regional languages were used in different parts of the world. A description is an account of the plant's habit, morphology and periodicity whereas a diagnosis is an author's definitive statement of the plant's diagnostic features, and circumscribes the limits outside which plants do not pertain to that named species. A diagnosis often states particular ways in which the species differs from another species of the same genus. Before the adoption of Latin as the accepted language of botanical nomenclature searching for names already in existence, and confirming their applicability to a particular plant, involved searching through literature in many languages. The requirement to use Latin was written into the rules by the International Botanical Congress in Vienna, in 1905. However, the American Society of Plant Taxonomists produced their own American Code of Nomenclature in 1947 and disregarded this requirement. Not until 1959 was international agreement achieved and then the requirement to use Latin was made retroactive to January 1st, 1935; the year of the Amsterdam meeting of the Congress.

The rules are considered at each International Botanical Congress, which normally occur at five-yearly intervals during peacetime. The International Code of Botanical Nomenclature (first published as such in 1952) was formulated at the Stockholm Congress of 1950. In 1930, the matter of determining the priority of specific epithets was the main point at issue. The practice of British botanists had been to regard the epithet which was first published after the

24

plant had been allocated to its correct genus as the correct name. This has been called the Kew Rule but it was defeated in favour of the rule which now gives priority to the epithet which was the first to be published (since May 1st, 1753 or, if Linnaeus had accepted it, earlier).

The 1959 International Botanical Congress in Montreal introduced the requirement under the Code that, for valid publication of a name, the author of the name (of a family or any taxon of lower rank) should cite a 'type' for that name and that this requirement should be retroactive to January 1st, 1958. The idea of a type goes back to A. P. de Candolle and it implies a representative collection of characteristics to which a name applies. The type in botany is a nomenclatural type; it is the type for the name and the name is permanently attached to it or associated with it. For the name of a family, the representative characteristics which that name implies are those embodied in one of its genera, which is called the type genus. In a similar way, the type for the name of a genus is the type species of that genus. For the name of a species or taxon of lower rank, the type is a specimen lodged in an herbarium or, in certain cases, published illustrations. The type need not, nor could it, be representative of the full range of entities to which the name is applied. Just as a genus, although having the features of its parent family, cannot be fully representative of all the genera belonging to that family, no single specimen can be representative of the full range of variety found within a species.

For a name to become the correct name for a plant or plant group, it must satisfy two sets of conditions. The first determines the name's legitimacy, that is, its accordance with the rules of name formation, and the second ensures proper publication. Publication has to be in printed matter which is distributed to the general public or, at least, to botanical institutions with libraries accessible to botanists generally. Since January 1st, 1953, this has excluded publication in newspapers and tradesmen's catalogues. As well as being effectively published, it must be validly published, with an accompanying description or diagnosis, an indication of its rank and of the nomenclatural type, and the author's name, as required by the rules. This publication requirement ensures that a date can be placed upon a name's publication and that it can therefore be properly considered in matters of priority.

The present scope of the Code is expressed in the principles, which have evolved from the de Candollean Lois:

1. Botanical nomenclature is independent of zoological nomenclature. The Code applies equally to names of taxonomic groups treated as plants whether or not these groups were originally so treated.
2. The application of names of taxonomic groups is determined by means of nomenclatural types.
3. The nomenclature of a taxonomic group is based upon priority of publication.
4. Each taxonomic group with a particular circumscription, position, and rank can bear only one correct name, the earliest that is in accordance with the rules, except in specified cases.
5. Scientific names of taxonomic groups are treated as Latin regardless of their derivation.
6. The rules of nomenclature are retroactive unless expressly limited.

The detailed rules are contained in the articles and recommendations of the Code and mastery of these can only be gained by practical experience.

There are still new species of plants to be discovered and an enormous amount of information yet to be sought for long-familiar species, particularly evidence of a chemical nature, and especially that concerned with proteins, which may provide reliable indications of phylogenetic relationships. For modern systematists, the greatest and most persistent problem is our ignorance about the apparently explosive appearance of a diverse array of flowering plants, some 100 million years ago, from one or more unknown ancestors. Modern systems of classification are still frameworks within which the authors arrange assemblages in sequences or clusters to represent their own idiosyncratic interpretation of the known facts. In addition to having no firm record of the early evolutionary pathways of the flowering plants, the systematist also has the major problems of identifying clear-cut boundaries between groups and of assessing the absolute ranking of groups. It is because of these continuing problems that, although the Code extends to taxa of all ranks, most of the rules are concerned with the names and naming of groups from the rank of family downwards.

Before moving on to the question of plant names at the generic and lower ranks, this is a suitable point at which to comment on the move towards the standardizing of the names of families in recent years. The new names for families are now starting to appear in books and catalogues, and some explanation in passing may help to dispel any confusion.

Each family can have only one correct name and that, of course, is the earliest legitimate one, except in cases of limitation of priority by conservation. In other words, there is provision in the Code for disregarding the requirement of priority when a special case is proved for a name to be conserved. Conservation of names is intended to avoid disadvantageous name changes, even though the name in question does not meet all the requirements of the Code. Names which have long-standing use and wide acceptability and are used in standard works of literature can be proposed for conservation and, when accepted need not be discarded in favour of a new and more correct name.

The names of families are plural adjectives used as nouns and are formed by adding the suffix *-aceae* to the stem, which is the name of an included genus. Thus, the buttercup genus *Ranunculus* gives us the name Ranunculaceae for the buttercup family and the water-lily genus *Nymphaea* gives us the name Nymphaeaceae for the water-lilies. A few family names are conserved, for the reasons given above, which do have generic names as their stems, although one, the Ebenaceae, has the name *Ebenus* Kuntze (1891) non Linnaeus (1753) as its stem. This is now called *Maba* and the name *Ebenus* L. is used for a genus of the pea family. There are eight families for which specific exceptions are provided and which can be referred to either by their long-standing, conserved names or, as is increasingly the case in recent floras and other published works on plants, by their names which are in agreement with the Code. These families, and their equivalents are:

Compositae	or Asteraceae (on the genus *Aster*)
Cruciferae	or Brassicaceae (on the genus *Brassica*)
Gramineae	or Poaceae (on the genus *Poa*)
Guttiferae	or Clusiaceae (on the genus *Clusia*)
Labiatae	or Lamiaceae (on the genus *Lamium*)
Leguminosae	or Fabaceae (on the genus *Faba*)
Palmae	or Arecaceae (on the genus *Areca*)
Umbelliferae	or Apiaceae (on the genus *Apium*)

Some botanists regard the Leguminosae as including three sub-families but others accept those three components as each having family status. In the latter case, the three families are the Caesalpiniaceae, the Mimosaceae and the Papilionaceae. The last of these has a name which refers to the pea- or bean-flowered adaptation to pollination by butterflies (*Papilio*) and which is not based upon the name of a plant genus. If a botanist wishes to retain the three-family concept, the name Papilionaceae is conserved against Leguminosae and the modern equivalent is Fabaceae. Consequently, the Fabaceae are either the entire aggregation of leguminous plant genera or that part of the aggregate which does not belong in either the Mimosaceae or the Caesalpiniaceae.

The name of a genus is a noun, or word treated as such,

and begins with a capital letter. It is singular and may be taken from any source whatever and may even be composed in an absolutely arbitrary manner. The etymology of generic names is, therefore, not always complete and, even though the derivation of some may be discovered, they lack meaning. By way of examples:

Portulaca from the Latin *porto* (I carry) *lac* (milk) translates as 'Milk carrier'.

Pittosporum from the Greek *pitto* (to tar) and *sporos* (a seed) translates as 'tar seed'.

Hebe was the goddess of youth and, amongst other things, the daughter of Jupiter. It cannot be translated further.

Petunia is taken from the Brazilian name for tobacco.

Tecoma is taken from a Mexican name.

Linnaea is one of the names which commemorate Linnaeus.

Sibara is an anagram of *Arabis*.

Aa is the name which Reichenbach gave to an orchid genus which he separated from *Altensteinia*. It has no meaning and, as others have observed, must always appear first in an alphabetic listing.

The name of a species is a binary combination of the generic name followed by a specific epithet. If the epithet is of two words they must be joined by a hyphen or united into one word. The epithet can be taken from any source whatever and may also be composed in an arbitrary manner. It would be reasonable to expect that the epithet should have a descriptive purpose, and there are many which do, but large numbers either refer to the native area in which the plant grows or commemorate a person (often the discoverer, the introducer into cultivation or a noble personage). The epithet may be adjectival (or descriptive), qualified in various ways with prefixes and suffixes, or a noun.

The name of a subdivision of a species is that of the species followed by a term donating the rank of the subdivision and an epithet formed in the same ways as the specific epithet.

The ranks concerned are *subspecies* (abbreviated to subsp. or ssp.), *varietas* (in English as variety, abbreviated to var.) and *forma* (in English as form, abbreviated to f.). When a subdivision of a species is named which does not include the nomenclatural type, it automatically establishes the equivalent subdivision which does contain the type and which, therefore, has the same epithet as the species itself. For example: *Veronica spicata* L. subsp. *hybrida* (L.) E. F. Warburg implies the existence of, and at once establishes, the typical *V. spicata* subsp. *spicata*. In this example, Warburg reduced Linnaeus' *V. hybrida* to a subspecies of *V. spicata*. Retention of Linnaeus' epithet and its attribition to him (in brackets) ensures that the erection of Warburg's subspecies also explains the disappearance of the former Linnaean species.

If all specific names were constructed in an arbitrary manner there would have been no enquiries of the writer and this book would not have been written. In fact, the etymology of plant names is a rich store of historical interest and conceals many facets of humanity ranging from the sarcasm of some authors to the humour of others. This is made possible by the wide scope available to authors for formulating names and because, whatever language is the source, names are treated as being in Latin. The large vocabulary of botanical Latin comes mostly from the Greek and Latin of ancient times but, since the ancients had few words which related specifically to plants and their parts, a Latin dictionary is of somewhat limited use in trying to decipher plant diagnoses. By way of examples, Table 1 gives the parts of the flower (illustrated in Fig. 1) and the classical words from which they are derived, together with their original sense.

The grammar of botanical Latin is very formal and much more simple that that of the classical language itself. Nevertheless, it is necessary to know that nouns (such as family and generic names) in Latin have gender, number and case and that the words which give some attribute to a noun

Table 1

Flower part	Greek	Latin	Former meaning
calyx	καλυξ	—	various kinds of covering
	καλιξ	—	cup or goblet
sepal	σκεπη		covering
corolla	—	*corolla*	garland or coronet
petal	πεταλον	—	leaf
stamen		*petalum*	metal plate
filament		*stamen*	thread, warp, string
anther		*filamentum*	thread
androecium	ανδροξ, οικοζ	*anthera*	potion of herbs
stigma	στιγμα		man, house
style	στυλοζ		tattoo or spot
carpel	καρπόζ	*stilus*	pillar or post
gynoecium	γυνη, οίκοζ		pointed writing tool
pistil	—	—	fruit
		pistillum	woman, house
			pestle

31

(as in adjectival specific epithets) must agree with the noun in each of these. Having gender means that all things (the names of which are called nouns) are either masculine or feminine or neuter. In English we treat almost everything as neuter, referring to nouns as 'it', except animals and most ships and aeroplanes (which are commonly held to be feminine). Gender is explained further below. Number means that things may be single (singular) or multiple (plural). In English we either have different words for the singular and plural (man and men, mouse and mice) or we convert the singular into the plural most commonly by adding an 's' (ship and ships, rat and rats) or more rarely by adding 'es' (box and boxes, fox and foxes) or rarer still by adding 'en' (ox and oxen). In Latin, the difference is expressed by changes in the endings of the words. Case is less easy to understand but it means the significance of the noun to the meaning of the sentence in which it is contained. It is also expressed by changes in the endings of the words. In the sentence 'The flower has charm', the flower is singular, is the subject of the sentence and has what is termed the nominative case. In the sentence 'I threw away the flower', I am now the subject and the flower has become the direct object with the accusative case. In the sentence 'I did not like the colour of the flower', I am again the subject, the colour is now the object and the flower has become a possessive noun and has the genitive case. In the sentence 'The flower fell to the ground', the flower is once again the subject (nominative) and ground has the dative case. If we add 'with a whisper', then whisper takes the ablative case. In other words, case confers on nouns an expression of their meaning in any sentence. This is shown by the ending of the Latin word, which changes with case and number, and in so doing changes the naked word into part of a sentence (Table 2).

Nouns fall into five groups, or declensions, as determined by their endings (Table 3). Genera are singular nouns and

Table 2

	Singular		Plural	
nominative	*flos*	the flower (subject)	*flores*	the flowers
accusative	*florem*	the flower (object)	*flores*	the flowers
genitive	*floris*	of the flower	*florum*	of the flowers
dative	*flori*	to or for the flower	*floribus*	to or for the flowers
ablative	*flore*	by, with or from the flower	*floribus*	by, with or from the flowers

Table 3

Declension	I	II		III				IV		V
Gender	f	m	n	m,f	n	m,f	n	m	n	f
Singular nom	-a	-us(er)	-um	*	*	-is(es)	-e(l)(r)	-us	-u	-es
acc	-am	-um	-um	-em	*	-em(im)	-e(l)(r)	-um	-u	-em
gen	-ae	-i	-i	-is	-is	-is	-is	-us	-us	-ei
dat	-ae	-o	-o	-i	-i	-i	-i	-ui(u)	-ui(u)	-ei
abl	-a	-o	-o	-e	-e	-i(e)	-i(e)	-u	-u	-e
Plural nom	-ae	-i	-a	-es	-a	-es	-ia	-us	-ua	-es
acc	-as	-os	-a	-es	-a	-es(is)	-ia	-us	-ua	-es
gen	-arum	-orum	-orum	-um	-um	-ium	-ium	-uum	-uum	-erum
dat	-is	-is	-is	-ibus	-ibus	-ibus	-ibus	-ibus	-ibus	-ebus
abl	-is	-is	-is	-ibus	-ibus	-ibus	-ibus	-ibus	-ibus	-ebus

* Denotes various irregular endings.

are treated as subjects, taking the nominative case. *Solanum* means 'the comforter' and derives from the use of night-shades as herbal sedatives. The gender of generic names is that of the original Greek or Latin noun or, if that was variable, is chosen by the author of the name. There are exceptions to this in which masculine names are treated as feminine, and fewer in which compound names which ought to be feminine are treated as masculine. As a general guide, names ending in *-us* are masculine unless they are trees (such as *Fagus*, *Pinus*, *Quercus*, *Sorbus*); names ending in *-a* are feminine and names ending in *-um*, are neuter; names ending in *-on* are masculine unless they can also take *-um*, when they are neuter, or the ending is *-dendron*, when they are also neuter (*Rhododendron* or *Rhododendrum*); names ending in *-ma* (as in terminations such as *-osma*) are neuter; names ending in *-is* are mostly feminine or masculine treated as feminine (*Orchis*) and those ending in *-e* are neuter; other feminine endings are *-ago*, *-odes*, *-oides*, *-ix*, and *-es*.

A recommendation for forming generic names to commemorate men or women is that these should all be treated as feminine, regardless of the sex of the person commemorated, and that the ending should be formed as follows:

for names ending in a vowel, terminate with *-a*

for names ending in *a*, terminate with *-ea*

for names ending with *ea*, do not change

for names ending in a consonant, add *-ia*

for names ending in *er*, add *-a*

for latinized names ending in *-us*, add *-ia*.

Generic names which are formed arbitrarily or derived from vernacular names have their ending selected by the author of the name

Epithets commemorating people

Specific epithets which are nouns are grammatically independent of the generic name; *Campanula trachelium* is literally 'little bell' (feminine) 'neck' (neuter). When they are derived from the names of people, they can either be used in the genitive case: *clusii* is the genitive singular of Clusius, the latinized version of l'Ecluse, and gives an epithet with the meaning 'of l'Ecluse'; or be treated as adjectives and then agreeing in gender with the generic noun: *Sorbus leyana* Wilmott is a tree taking the feminine gender, despite the masculine ending, and so the epithet which commemorates Augustin Ley also takes the feminine ending. The epithets are formed by:

adding '*-i*' to names ending in 'er' or a vowel other than '-a' or

adding '*-e*' to names ending in 'a' or

adding '*-ii*' (masculine) or '*-iae*' (feminine) to names ending with a consonant or, for the respective cases in the plural

adding '*-orus*' (masculine) or '*-arus*' (feminine) or

adding '*-rus*' or

adding '*-iorum*' (masculine) or '*-iarum*' (feminine).

Adjectival epithets are formed by:

adding '*-ianus -a -um*' to names ending in a consonant or

adding '*-anus -a -um*' to names ending in a vowel other than '-a' or

adding '*-nus -a um*' to names ending in 'a'.

Names commemorating women are treated as feminine in the genitive case.

Geographical epithets

When an epithet is derived from the name of a place, usually to indicate the plant's native area but also sometimes to indicate the area or place from which the plant was first known or in which it was produced horticulturally, it is adjectival and takes one of the endings *-ensis*, *-(a)nus*, *-inus*, *-ianus* or *-icus*. Geographical epithets are sometimes inaccurate because the author of the name was in error as to the true origin of the plant, or obscure because the ancient classical names are no longer familiar to us. As with epithets

Table 4

Masculine	Feminine	Neuter		
-us	-a	-um	*hirsutus*	(hairy)
-is	-is	-e	*brevis*	(short)
-os	-os	-on	*acaulos*	(stemless)
-er	-era	-erum	*asper*	(rough)
-er	-ra	-rum	*scaber*	(rough)
-ax	-ax	-ax	*fallax*	(false)
-ex	-ex	-ex	*duplex*	(double)
-ox	-ox	-ox	*ferox*	(very prickly)
-ans	-ans	-ans	*reptans*	(creeping)
-ens	-ens	-ens	*repens*	(creeping)
-or	-or	-or	*tricolor*	(three-coloured)
-oides	-oides	-oides	*bryoides*	(moss like)

Table 5

Masculine	Feminine	Neuter		
-us	-a	-um	*longus*	(long)
-ior	-ior	-ius		(longer)
-issimus	-issima	-issimum		(longest)
-is	-is	-e	*gracilis*	(slender)
-ior	-ior	-ius		(slenderer)
-limus	-lima	-limum		(slenderest)
-er	-era	-erum	*tener*	(thin)
-erior	-erior	-erius		(thinner)
-errimus	-errima	-errimum		(thinnest)

which are derived from proper names to commemorate people, or from former generic names or vernacular names which are treated as being Latin, it is now customary to start them with a small initial letter but it remains permissible to give them a capital initial.

It will be clear that because descriptive, adjectival epithets must agree with the generic name, the endings must change in gender, case and number: *Dipsacus fullonum* L. has the generic name used by Dioscorides meaning 'dropsy', alluding to the accumulation of water in the leaf-bases, and an epithet which is the masculine genitive plural of *fullo*, a fuller, and which identifies the typical form of this teasel as the one which was used to clean and comb up a 'nap' on cloth. The majority of adjectival epithet endings are as in the first two of the examples in Table 4.

Comparative epithets are informative because they provide us with an indication of how the species contrasts with the general features of the other members of the genus (Table 5).

Hybrids

Hybrids are particularly important as cultivated plants but are also a feature of many plant groups in the wild, especially woody perennials such as willows. The rules for the names and naming of hybrids are contained in the Botanical Code but are equally applicable to cultivated plant hybrids.

For the name of a hybrid between parents from two different genera, a name can be constructed from the two generic names, in part or in their entirety, as a condensed formula; × *Mahoberberis* is the name for hybrids between the genera *Mahonia* and *Berberis* (in this case the cross is only bigeneric because the name *Mahonia* is conserved against *Berberis*) and × *Fatshedera* is the name for hybrids between the genera *Fatsia* and *Hedera*. Alternatively a formula can be used in which the names of the parent genera are linked by the sign for hybridity ' × '; *Mahonia* × *Berberis* and *Fatsia* × *Hedera*. Hybrids between parents from three genera are also named either by a formula or a condensed formula and, in all cases, the condensed formula is treated as a generic name if it is published with a statement of parentage. When published, it becomes the correct generic name for any hybrids between species of the named parental genera. A

38

third alternative is to construct a commemorative name in honour of a notable person and to end it with the termination *-ara*; *Sanderara* is the name applied to the orchid hybrid *Brassia* × *Cochlioda* × *Odontoglossum* and commemorates H. F. C. Sander, the British orchidologist.

A name formulated to define a hybrid between two particular species from different genera can take the form of a species name, and then applies to all hybrids produced subsequently from those parent species: × *Fatshedera lizei* Guillaumin is the name for hybrids between *Fatsia japonica* (Thunb.) Decne. & Planch. and *Hedera helix* L. 'Hibernica' and × *Cupressocyparis leylandii* (Jackson & Dallimore) Dallimore is the name for hybrids between *Chamaecyparis nootkatensis* (D. Don) Spach and *Cupressus macrocarpa* Hartweg ex Godron. Because the parents themselves are variable, the progeny of repeated crosses may be distinctive and deserving of naming as 'nothomorphs': × *Cupressocyparis leylandii* nm. 'Naylor's Blue'. Nothomorphs are now treated as cultivars (varieties) and the hybridity of any taxon (nothotaxon) can be specified by adding 'notho-' or 'n' to the term denoting its rank, as an alternative to use of the ' × ', so that *Sanderara* is a nothogenus and *Cupressocyparis leylandii* is a nothospecies.

Hybrids between species in the same genus are also named by a formula or by a new distinctive epithet; *Digitalis lutea* L. × *D. purpurea* L. and *Nepeta* × *faassenii* Bergmans ex Stearn are both correct names for hybrids. In the example of *Digitalis*, the order in which the parents are presented happens to be the correct order, with the seed parent first. It is permissible to indicate the roles of the parents by including the symbols for female '♀' and male '♂', when this information is known, or otherwise to present the parents in alphabetical order.

Certain hybrids generate fertile offspring as a consequence of their chromosome complements having been doubled. Since such doubling permits them to behave as quite normal sexual species, they are named as such. Thus, the tertaploid (with four, instead of the normal two sets of chromosomes) of the hybrid between *Digitalis grandiflora* Mill. and *D. purpurea* L. has been named as *Digitalis mertonensis* Buxton & Darlington, and does not require the ' × ' sign to indicate its nothospecific nature.

The International Code of Nomenclature for Cultivated Plants

In 1952, the Committee for the Nomenclature of Cultivated Plants of the International Botanical Congress and the International Horticultural Congress in London adopted the International Code of Nomenclature for Cultivated Plants. Sometimes known as the Cultivated Code, it was first published in 1953 and has been revised at intervals since then. This Code formally introduced the term 'cultivar' to encompass all varieties or derivatives of wild plants which are raised under cultivation. Its aim is to 'promote uniformity and fixity in the naming of agricultural, sylvicultural and horticultural cultivars (varieties)'. There can be no doubt that the diverse approaches to naming garden plants, by common names, by botanical names, by mixtures of botanical and common names, by group names and by fancy names, is no less complex than the former unregulated use of common or vernacular names.

This Code governs the names of all plants which retain their distinctive characters when reproduced sexually (by seed) or vegetatively in cultivation. Because the Code does not have legal status, the commercial interests of plant breeders are guarded by the Council of the International Union for the Protection of New Varieties of Plants (U.P.O.V.). In Britain, the Plant Varieties Rights Office works with the Government to have U.P.O.V.'s guidelines implemented.

The Cultivated Code accepts the rules of botanical nomenclature and the retention of those names for plants taken into cultivation from the wild. It recognizes only the one category of garden-maintained variant, the cultivar or garden variety (cv.), which should not be confused with the botanical *varietas*. Unlike wild plants, cultivated plants receive unnatural treatment and selection pressures from man and are maintained by him. The term cultivar covers:

clones which are derived vegetatively from a single parent,

lines of selfed or inbred individuals,

series of crossbred individuals, and

assemblages of individuals which are resynthesized only by cross-breeding (e.g. F_1 hybrids).

Since January 1st, 1959, the names of cultivars have had to be 'fancy names' in common language and not in Latin. Fancy names come from any source. They can commemorate anyone, not only persons connected with botany or plants, or they can identify the nursery of their origin, or be descriptive, or be truly fanciful. Those which had Latin garden variety names were allowed to remain in use; *Nigella damascena* L. has the old varietal names *alba* and *flore pleno* and also has the modern cultivar with the fancy name 'Miss Jekyll'. In the glossary, no attempt has been made to include fancy names but a few of the earlier Latin ones have been included.

In order to be distinguishable, the fancy names have to be printed in a type-face unlike that of the species name and to be given capital initials. They also have to be either placed between single quotation marks, as above, or be preceded by 'cv.'. Thus, *Salix caprea* L. 'Kilmarnock', or cv. Kilmarnock, is a weeping variety of the goat willow and is also part of the older variety *S. caprea* var. *pendula*; other examples are *Geranium ibericum* Cav. 'Album' and *Acer davidii* Franchet cv. George Forrest.

Fancy names can be applied to an unambiguous common name, such as potato 'Duke of York' for *Solanum tuberosum* L. cv. Duke of York, or to a generic name such as *Cucurbita* 'Table Queen' for *Cucurbita pepo* L. cv. Table Queen, or of course to the botanical name, even when this is below the level of species, *Rosa sericea* var. *omiensis* 'Praecox'. However, the same fancy name may not be used twice within a group (cultivar class) if such duplication would cause ambiguity, Thus, cherries and plums are in distinct cultivar classes and we would never refer to either by the generic name, *Prunus*, alone. Consequently, the same fancy name could be used for a cultivar of a cherry and for a cultivar of a plum.

For some extensively bred crops and decorative plants there is a long-standing supplementary category, the group. By naming the group in such plants, a greater degree of accuracy is given to the garden name; such as pea (wrinkle-seeded group) 'Laxton's Progress', and *Rosa* (rambler) 'Albéric Barbier' and *Rosa* (rugosa) 'Agnes'.

To ensure that a cultivar has only one correct name, the Cultivated Code requires that priority acts and, to achieve this, publication and registration are necessary. As with botanical names, cultivars can have synonyms and the problems are increased in this respect because it is permissible to translate the fancy names into other languages. To establish a fancy name, publication has to be by printed matter which is dated and distributed to the public. For the more popular groups of plants, usually genera, there are societies which maintain statutory registers of names and the plant breeding industry has available to it the Plant Variety Rights Office as a statutory registration body for crop-plant names. In some countries it is possible to register cultivar names as trade marks for commercial protection, including patent rights on vegetatively propagated cultivars.

One group of plants which is almost entirely within the province of gardening is that of the graft chimaeras, or graft hybrids. These are plants in which a mosaic of tissues from the two parents in a grafting partnership results in an individual plant upon which shoots resembling each of the parents, and in some cases shoots of intermediate character, are produced in an unpredictable manner. Unlike sexually produced hybrids, the admixture of the two parents' contributions is not at the level of the nucleus in each and every cell but is more like a marbling of a ground-tissue of one parent with streaks of tissue of the other parent. Chimaeras can also result from mutation in a growing point, from which organs are formed composed of normal and mutant tissues. In all cases, three categories may be recognized, called sectorial, mericlinal and periclinal chimaeras, in terms of the extent of tissue 'marbling'. The chimaeral condition is denoted in plant names by the addition sign ' + ', instead of the ' × ' used for true hybrids. A chimaera which is still fairly common in Britain is that named + *Laburnocytisus adamii* C. K. Schneider. This was the result of a graft between *Cytisus purpureus* Scop. and *Cytisus laburnum* L., which are now known as *Chamaecytisus purpureus* (Scop.) Link. and *Laburnum anagyroides* Medicus, respectively. Although its former name, *Cytisus + adamii* would not now be correct, the name *Laburnocytisus* meets the requirement of combining substantial parts of the parental generic names, and can stand.

Botanical terminology

There is nothing accidental about the fact that in our everyday lives we communicate at two distinct levels. Our 'ordinary' conversation employs a rich, dynamic language in which meaning can differ from one locality to another and change from time to time. Our 'ordinary' reading is of a written language of enormous diversity – ranging from contemporary magazines which are intentionally erosive of good standards, to the high-quality prose of serious writers. However, when communication relates to specific topics, in which ambiguity is an anathema, the language which we adopt is one in which 'terminology' is relied upon to convey information accurately and incontroversially. Thus, legal, medical and all scientific communications employ terms which have widely accepted meanings and which therefore convey those meanings in the most direct way. Because, like the botanical terms for the parts of the flower, these terms are derived predominantly from classical roots and have long-standing acceptance, they have the added advantage of international currency.

This glossary contains many examples of words which are part of botanical terminology as well as being employed as descriptive elements of plant names. Such is the wealth of botanical terminology that an attempt here to discriminate between and explain all the terms which relate, say, to the surface of plant leaves and the structures (hairs, glands and deposits) which subscribe to that texture would make tedious reading. However, terms which refer to such conspicuous attributes as leaf shape and the form of inflorescences are very commonly used in plant names and, since unambiguous definition would be lengthy, are illustrated as figures.

More extensive glossaries of terminology can be found in textbooks and floras but the sixth edition of Willis' *Dictionary of Flowering Plants and Ferns* (1955) is a particularly rewarding source of information.

The glossary

The glossary is for use in finding out the meanings of the names of plants. There are many plant names which cannot be interpreted or which yield very uninformative translations. In general, commemorative and geographical names and anagrams have been omitted from the glossary and those which are included are generally the ones which might be confused with names of classical derivation. In certain groups such as garden plants from, say, China and exotics such as the profuse orchid family, commemorative names have been applied to plants more frequently than in most other groups. A number of the more significant of these commemorative names has been included in the glossary and if the reader wishes to add further significance to such names, he will find it mostly in literature on plant-hunting and hybridization.

Generic names in the European flora are mostly of ancient origin. Their meanings, even of those which are not taken from mythological sources, are seldom clear and many have had their applications changed and are now used as specific epithets. Generic names of plants discovered throughout the world in recent times have mostly been constructed to be descriptive and will yield to translation. The glossary does not attempt to list large numbers of generic names as these can be found elsewhere (Farr, 1980), but does permit translation of those which are descriptive.

As an example of how the glossary can be used, we can consider the name *Sarcococca ruscifolia*. This is the name given by Stapf to plants which belong to Lindley's genus

Sarcococca of the family Buxaceae, the box family. In the glossary we find *sarc-*, *sarco-* meaning fleshy- and *-coccus -a -um* meaning -berried and from this we conclude that *Sarcococca* means fleshy-berry (the generic name being a singular noun). We also find *rusci-* meaning *Ruscus*-like or resembling butcher's broom and *-folius -a -um* meaning -leaved and we conclude that this species of fleshy-berry has leaves which resemble the prickly cladodes (leaf-like branches) of *Ruscus*. The significance of this generic name lies in the fact that dry fruits are more typical in the box family than fleshy ones.

From this example, it should be clear that names can be constructed from adjectives or adjectival nouns to which prefixes or suffixes can be added, thus giving them further qualification. As a general rule, epithets which are formed in this way have an acceptable interpretation when '-ed' is added to the English translation; this would render *ruscifolia* as *Ruscus*-leaved.

It will be noted that *Sarcococca* has a feminine ending '-a' and that *ruscifolia* takes the same gender. However, if the generic name had been of the masculine gender the epithet would have become *ruscifolius* and if of the neuter gender then it would have become *ruscifolium*. For this reason the entries in the glossary are given all three endings which, as pointed out earlier, mostly take the form *-us*, *-a*, *-um* or *-is*, *-is*, *-e*.

Where there is the possibility that a prefix which is listed could lead to the incorrect translation of some epithet, the epithet in question is listed close to the prefix and to an example of an epithet in which the prefix is employed. Examples are:

   aer-, meaning air- or mist-, gives *aerius a -um* meaning airy, or lofty.

   aeratus -a -um, however, means bronzed (classically made of bronze).

   caeno-, from the Greek *cainos*, means fresh-.

47

caenosus -a -um, is from the Latin *caenum* which means of mud or filth.

Many of the epithets which may cause confusion are either classical geographic names or terms which retain a more specific meaning of the classical language. Many more such epithets exist than are listed in this glossary.

The glossary

a-, ab- away from, downwards, without, very
ac-, ad-, af-, ag-, al-, an-, ap-, ar-, as-, at- near-, towards-
abbreviatus -a -um shortened
Abies rising one (tall trees)
-abilis -is -e -able, -capable of (preceded by some action)
abortivus -a -um with missing or malformed parts
abro delicate
abrotonoides *Artemisia*-like (from an ancient Greek name for
     wormwood or mugwort)
abruptus -a -um ending suddenly
abscissus -a -um cut off
absinthius -a -um from an ancient Greek or Syrian name for
     wormwood
abyssinicus -a -um of Abyssinia, Abyssinian
Acaena Thorny one
acantho- spiny-, thorny-
acaulis -is -e, acaulon, acaulos lacking an obvious stem
Acer sharp (either from its use for lances or its leaves)
acer, acris, acre sharp-tasted, acid
aceras without a spur, not horned
acerbus -a -um harsh-tasted
aceroides maple-like
acerosus -a -um pointed, needle-like
acetabulosus -a -um saucer-shaped
acetosus -a -um acid, sour
acetosellus -a -um slightly acid
-aceus -a -um -resembling (preceded by a plant name)
Achillea after the Greek warrior Achilles
acicularis -is -e needle-shaped
aciculus -a -um sharply pointed (e.g. leaf-tips)
acidosus -a -um, acidus -a -um acid, sharp, sour
acinaceus -a -um, acinaciformis -is -e scimitar-shaped
acinifolius -a -um *Acinos*-leaved, basil-thyme-leaved

49

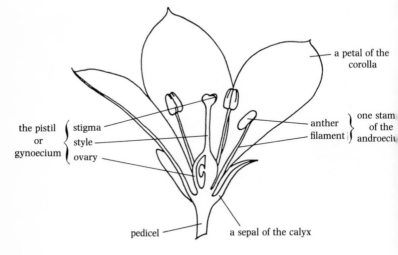

Fig. 1. The parts of a flower as seen in a stylized flower which is cut vertically in half.

acmo- pointed- (followed by a part of a plant)

Aconitum the name used by Theophrastus (poisonous)

Acorus Dioscorides' name for an iris, without a pupil (its use for cataract)

acro- towards the top-, highest- (followed by a verb e.g. fruiting, or a noun e.g. hair)

Actaea from the Greek name for elder (the shape of the leaves)

actino- radiating- (followed by a part of a plant)

aculeatus -a -um having prickles, thorny, prickly

aculeolatus -a -um having small prickles or thorns

acuminatus -a -um with a long, narrow and pointed tip (see Fig. 7(*c*)

acutus -a -um, acuti- acutely pointed, sharply angled at the tip

adamantinus -a -um from Diamond Lake, Oregon, U.S.A.

aden-, adeno-, gland-, glandular-

adenotrichus -a -um glandular-hairy

Adiantum an ancient Greek name, unwetted (remains unwetted under water)

admirabilis -is -e to be admired

adnatus -a -um joined together

Adonis (see *Anemone*)

Adoxa without glory

adpressus -a -um pressed together, lying flat against (e.g. the hairs on a stem)

adscendens curving up from a prostrate base, half-erect

adsurgens rising up

adulterinus -a -um of adultery (intermediacy suggesting hybridity)

aduncus -a -um hooked, having hooks

Aegopodium goat's foot (the leaf shape)

aemulus -a -um imitating, rivalling

aeneus -a -um bronzed

aequalis -is -e, aequali- equal, equally-

aequinoctialis -is -e of the equinox (the flowering time)

aer- air-, mist-

aeratus -a -um bronzed

aerius -a -um airy, lofty

aeruginosus -a -um rusty, verdigris-coloured

aeschyno- shy-, to be ashamed-

aestivalis -is -e of summer

aestivus -a -um developing in summer

aethiopicus -a -um of Africa, African, of N. E. Africa

Aethusa burn (pungency)

-aeus belonging to (a place)

afer, *afra*, *afrum* African

affinis -is -e related, similar to

Agapanthus love-flower

agastus -a -um charming, pleasing

Agave admirable

agetus -a -um wonderful

agglomeratus -a -um in a close head, congregated together

agglutinatus -a -um glued or joined together

aggregatus -a -um clustered together

-ago -like

agrarius -a -um, *agrestis -is -e* of fields, wild on arable land

Agrimonia cataract (medicinal use)

Agropyrum field wheat

Agrostis field

Ailanthus a Moluccan name signifying 'reaching to Heaven'

Aira old Greek name for darnel grass

Aizoon always alive

ajacis -is -e of Ajax (whose blood produced a flower marked with AIA)

Ajuga corrupted Latin for abortifant

alabastrinus -a -um like alabaster

alatus -a -um, *alati-*, *alato-* winged, with protruding ridges which are wider than thick

albatus -a -um turning white

albens white

albescens turning white

albicans whitish

albidus -a -um, *albido-* whitish

albus, *-a -um*, *albi-*, *albo-* dead-white

alceus -a -um mallow-like, from 'alcea' the name used by Dioscorides

Alchemilla from Arabic reference either to reputed magical properties or to fringed leaves of some species

alcicornis -is -e elk-horned

aleppicus -a -um of Aleppo, N. Syria

aleur-, *aleuro-* mealy-, flowery- (surface texture)

aleuticus -a -um Aleutian

algidus -a, *-um* cold, of high mountains

-alis -is -e -belonging to

alkekengi a name used by Dioscorides

allantoides sausage-shaped

allatus -a -um introduced, not native

alliaceus -a -um, allioides *Allium*-like, smelling of garlic
Alliaria garlic-smelling
allionii for Carlo Allioni (author of *Flora Pedemontana*)
Allium the ancient Latin name for garlic
allo- diverse-, several-, different-, other-
alni- alder-, alder-like-
Alopercurus fox's tail
alpester -ris -re of mountains
alpicolus -a -um of high mountains
alpinus -a -um alpine, of mountain pastures
alsaticus -a -um from Alsace, France
Alsine a named used by Dioscorides for a chickweed-like plant
also- leafy-, of groves-
alternans alternating
alternatus -a -um alternate, not opposite
Althaea a name used by Theophrastus, healer
alti-, alto-, altus -a -um tall, high
altilis -is -e nutritious
alutaceus -a -um of the texture of soft leather
alveolatus -a -um with shallow pits, alveolar
Alyssum pacifier, not rage, without fury
amabilis -is -e pleasing
amaranticolor purple
amarellus -a -um, amarus -a -um bitter (as in the amaras or
  bitters of the drinks industry)
amaurus -a -um dark
amb-, ambi- around
ambigens, ambiguus -a -um doubtful, of uncertain relationship
ambly- blunt
amboinensis -is -e from Amboina, Indonesia
ambrosia elixir of life, food of the gods
amentaceus -a -um having catkins
amethystea the colour of amethyst gems
amicorum of the Friendly Isles, Tongan
amictus -a -um clad, clothed
amiculatus -a -um cloaked, mantled
Ammi a name used by Dioscorides, sand
ammophillus -a -um sand-loving (the habitat)
amoenus -a -um pleasing, delightful
amomum purifying (*Amomum* was used to cure poisoning)
ampelo- wine-
amphibius -a -um with a double life, growing both on land and in
  water

amphi-, *ampho-* on both sides-, in two ways-, both-, double-

amplexicaulis -is -e embracing the stem (e.g. base of the leaf, see Fig. 6(*d*))

amplus -a -um large

amplissimus -a -um very large

ampullaceus -a -um, *ampullaris -is -e* flask-shaped

amygdalinus -a -um of almonds, almond-like

an-, *ana-* up-, upon-, upwards-, above-, without-, backwards-, again-

Anacamptis bent back (the long spur of the flower)

Anagallis unpretentious, without boasting, without adornment

anagyroides resembling *Anagyris*, curved backwards

anastaticus -a -um rising up, resurrection plant

anatolicus -a -um from Anatolia, Turkish

anceps two-edged (stems)

ancistro- fish-hook-

Anchusa strangler (astringent properties)

ancylo- hooked-

andegavensis -is -e from Angers in Anjou, France

andro-, *andrus -a -um* male, stamened, anthered

androgynus -a -um with male and female flowers on the same head

Andromeda after the Ethiopean princess rescued by Perseus from the sea monster

Androsaemum man's blood (the blood-coloured juice of the berries)

Anemone a name used by Theophrastus. Possibly a corruption of Na'mán, semitic name for Adonis whose blood gave rise to the crimson *Anemone coronaria*. It could also be a corruption of an invocation to the goddess of retribution, Nemesis. Commonly called windflower

anfractuosus -a -um twisted, bent

Angelica from the Latin for an angel

angio- urn-, vessel-, (enclosed-)

anglicus -a -um, *anglicorum* English

anguinus -a -um serpentine, snake-like-shaped

angularis -is -e angular

anguligerus -a -um having hooks

angulosus -a -um having angles

angusti-, *angustus -a -um* narrow

Anisantha unequal-flowered (flowers vary in sexuality)

anisatus -a -um aniseed-scented

aniso- unequal-, unequally-, uneven-

ankylo- crooked-

annotinus -a -um one year old, of last year (with distinct annual increments)

annularis -is -e ring-shaped

annulatus -a -um having rings

annuus -a -um annual

ano- upwards-, up-

anomalus -a -um unlike its allies

anosmus -a -um without fragrance, scentless

ansatus -a -um, ansiferus -a -um having a handle

anserinus -a um of the goose, of goose meadows

ante- before-

Antennaria feeler (the hairs of the pappus)

anthelmenthicus -a -um vermifuge, worm-expelling

Anthemis flowery

-anthemus -a -um, -anthes -flowered

antherotes brilliant

antho- flower-

anthora like *Ranunculus thora* in poisonous properties

Anthoxanthum yellow flower

anthracinus -a -um coal-black

anthropophagorus -a -um of the man-eaters

anthropophorus -a -um man-bearing (flowers of man orchid)

Anthurium flower tail (the long spadix)

-anthus -a -um -flowered

Anthyllis downy flower (hair on calyx)

anti- against-, opposite, opposite to-, like-

antidysentericus -a, -um of dysentery (use as a medical treatment)

antillarus -a -um from the Antilles, West Indies

antipodus -a -um of the Antipodes

antipyreticus -a -um against fire (the moss *Fontinalis antipyretica* was packed around chimneys to prevent thatch from igniting)

antiquorus -a -um of the ancients

antiquus -a -um ancient

Antirrhinum nose-like

anulatus -a -um ringed, with rings

-anus -a -um -belonging to, -having

anvegadensis -is -e see *andegavensis -is -e*

ap-, apo- without-, away from-, downwards-

aparine a name used by Theophrastus (clinging, seizing)

Apera a named used by Adanson (meaning uncertain)

apertus -a -um open, bare, naked

aphaca a name used in Pliny for a lentil-like plant

aphyllus -a -um without leaves, leafless (perhaps at flowering time)

Apium a name used in Pliny for a celery-like plant. Some relate it to the Celtic 'apon', water, as its preferred habitat

apiatus -a -um bee-like, spotted

apicatus -a -um with a pointed tip

apiculatus -a -um with a small broad point at the tip, apiculate (see Fig. 7(*e*))

apifer -era -erum bee-like, bee-bearing (flowers of bee orchid)

apodectus -a -um acceptable

apodus -a -um without a foot, stalkless

appendiculatus -a -um with appendages

applanatus -a -um flattened out

appressus -a -um lying close together, adpressed

approprinquatus -a -um near, approaching (resemblance to another species)

apricus -a -um sun-loving

apterus -a -um without wings, wingless

aquaticus -a -um growing in water

aquatilis -is -e growing under water

aquifolius -a -um with pointed leaves, spiny-leaved

Aquilegia eagle (claw-like nectaries)

aquilinus -a -um of eagles, eagle-like (the appearance of the vasculature in a cut rhizome)

aquilus -a -um blackish-brown

arabicus -a -um, *arabus -a -um* of Arabia, Arabian

Arabis Arabian, of Arabia

arachnites spider-like

arachnoideus -a -um cobwebbed, covered with a fine weft of hairs

araneosus -a -um spider-like, cobwebby

aranifer -era -erum spider-bearing

araucanus -a -um from the name of a tribe of Chilean Indians

Araucaria the Chilean Indian name

arboreus -a -um tree-like

arbusculus -a -um shrubby, small-tree-like

archi- primitive, beginning

arct-, *arcto-* bear-, northern-

Arctium bear (the shaggy hair)

arcturus -a -um bear's-tail-like

arcuatus -a -um bowed, curved

arenarius -a -um, *arenicola*, *arenosus -a -um* of sandy places, growing in sand

arenastrus -a -um resembling *Arenaria*

areolatus with distinct angular spaces (in the leaves)
Argemone a name used by Dioscorides for another plant
(medicinal use as a remedy for cataract)
argenteus -a -um silvery
argillaceus -a -um whitish
argo- pure white-
argutus -a -um sharply toothed or knotched
argyreus -a -um, argyro- silver-, silvery
aria a name used by Theophrastus for a Whitebeam
aridus -a -um dry, of dry habitats
arietinus -a -um ram-horned
-aris -is -e -pertaining to
aristatus -a -um with a beard, awned, aristate (see Fig. 7(*g*))
Aristolochia birth improver (abortifacient property)
-arius -a -um -belonging to, -having
arizelus -a -um notable
Arnoseris lamb succour
armatus -a -um armed, thorny
armeniacus -a -um apricot-yellow, from Armenia (mistake for
China)
armenus -a -um of Armenia, Armenian
Armeria ancient Latin name for a *Dianthus*
armillatus -a -um bracelet-like
arnoldianus -a -um of the Arnold Arboretum, Massachusetts
aromaticus -a -um fragrant
arrectus -a -um raised up, erect
arrhen-, arrhena- male-, stamen-
Arrhenatherum male awn (the lower spikelet is male and
awned)
arrhizus -a -um lacking roots, rootless
Artemisia after Queen Artemisia of Caria, Asia Minor
articulatus -a -um, arthro-, arto- joint-, jointed-
Arundo old Latin name
Arum a name used by Theophrastus
arundinaceus -a -um reed-like
arvalis -is -e of arable land, of cultivated land, from Arvas,
N. Spain
arvensis -is -e of the field, of ploughed fields
Asarum a name used by Dioscorides
ascendens upwards, ascending
-ascens -becoming, -turning to
asco- bag-
asper -era -erum rough (surface texture)

57

aspermus -a -um seedless, without seed
aspernatus -a -um despised
aspersus -a -um sprinkled
Aspidium shield (shape of indusium)
Asplenium spleen (medicinal use of)
assimilis -is -e resembling, like, similar to
assurgens, assurgenti- rising upwards, ascending
Aster star
-aster, -era -erum somewhat resembling
asterias star-like
asterioides aster-like
asthmaticus -a -um of asthma (medicinal use for)
astictus -a -um unspotted, immaculate
Astilbe without brilliance (the flowers)
Astragalus name used by Pliny for a plant with vertebra-like
    knotted roots
astro- star-shaped
ater, atra, atrum mat-black
athamanticus -a -um, athemanticus -a -um of Mount Athamas,
    Sicily
Athanasia immortal, without death (funerary use)
athero- bristle-
-aticus -a -um, -atilis -is -e -from (a place)
atlanticus -a -um of the Lesser Atlas Mountains, North Africa
atratus -a -um blackish, clothed in black
atri-, atro- very dark- (a colour)
Atropa inflexible (one of the goddesses of fate)
atrovirens very dark green
-atus -a -um -like, -having
aucuparius -a -um bird-catching, of bird catchers (fruit used as
    bait)
augustus -a -um stately, noble, tall
aulicus -a -um courtly
aulo- tube-, furrowed-
aurantiacus -a -um, aurantius -a -um orange-coloured
aurarius -a -um, auratus -a -um golden, ornamented with gold
aurelianensis -ia -e from Orleans, France
aureo-, aureus -a -um golden-yellow
auricomus -a -um with golden hair
auriculatus -a -um lobed like an ear, with ear-like lobes
aurigeranus -a -um from Ariège, France
auritus -a -um with ears, long-eared
aurorius -a -um orange

aurosus -a -um golden

australasiacus -a -um, australiensis -is -a Australian

australis -is -e southern

austriacus -a -um Austrian, from Austria

autumnalis -is -e of the autumn (flowering or growing)

avellanus -a -um hazy, from Avella, Italy

avenaceus -a -um oat-like

avicularis -is -e of small birds, eaten by small birds

avius -a -um of birds

Azalea formerly the name for *Loiseleuria*, of dry habitats

Azolla thought to refer to its inability to survive out of water

azureus -a -um sky-blue

baccatus -a -um having berries, fruits with fleshy or pulpy coats

baccifer -era -erum bearing berries

bacillaris -is -e staff-like, stick-like

badius -a -um reddish-brown

baeticus -a -um from Spain (Baetica), Andalusian

balansae for Benedict Balansa, a French plant collector

balanus the ancient name of an acorn

baldenis -is -e the Mount Baldo, N. Italy

baldschuanicus -a -um from Baldschuan, Bokhara

balsamifer -era -erum yielding a balsam, producing a fragrant
  resin

banaticus -a -um from Banat, Romania

Baphia dye (cam-wood gives a red dye and is also used for violin
  bows)

baphicantus -a -um of the dyers, dyers'

barba-jovis Jupiter's beard

Barbarea after St Barbara

barbarus -a -um foreign, from Barbary, North African coast

barbatus -a -um with tufts of hair, bearded

barbellatus -a -um having small barbs

barbigerus -a -um bearded

barometz from a Tartar word meaning lamb (the woolly fern's
  rootstock)

bary- heavy-

basi- of the base-, from the base-

basilaris -is -a relating to the base

basilicus -a -um princely, royal

batatas Haitian name for sweet potato

batrachioides water buttercup-like

Batrachium froggy (amphibious nature of water buttercup)

beccabunga from an old German name 'Bachbungen', mouth-smart or streamlet-blocker

belladonna beautiful lady, the juice of deadly nightshade was used to beautify by dilating the eyes when introduced as drops

bellatulus -a -um, bellulus -a -um somewhat beautiful

bellidiformis -is -e, belloides daisy-like

Bellis a name used in Pliny, pretty

bellus -a -um beautiful

benedictus -a -um well spoken of, blessed

benjamina from an Indian name

benzoin from an Arabic name

Berberis from the Arabic for barbary (North African)

Beta the Latin name for beet

betaceus -a -um beet-like

Betonica from a name in Pliny for a medicinal plant from Vettones in Spain

betonicifolius -a -um betony-leaved

Betula pitch (bitumen is distilled from the bark)

betulinus -a -um, betuloides, bętulus -a -um birch-like

bi-, bis- two-, twice-

bicolor of two colours

Bidens two teeth (the scales on the fruit's apex)

biennis -is -e lasting for two years, biennial

bifidus -a -um deeply two-cleft, bifid

-bilis -is -e -able, -capable

Biophytum life plant (sensitive leaves)

biserratus -a -um twice-toothed, double-toothed (leaf margin teeth themselves toothed)

bistortus -a -um twice-twisted (the roots)

bithynicus -a -um from Bithynia, Asia Minor

bituminosus -a -um tarry, clammy, adhesive

blandus -a -um pleasing, charming, not harsh, bland

blattarius -a -um an ancient Latin name, cockroach-like

blepharo- fringe-, eyelash-

blepharophyllus -a -um with fringed leaves

boeoticus -a -um from Boeotia, near Athena, Greece

bombyci- silk-

bombycinus -a -um silky

bona-nox good night (night flowering)

bonariensis -is -e from Buenos Aires, Argentina

bonduc an Arabic name for a hazel-nut

bononiensis -is -e from Bologna, N. Italy

bonus-henricus good King Henry (allgood or mercury)

Boophone ox-killer (narcotic property)

borbonicus -a -um from Réunion Island, Indian Ocean, or for the French Bourbon kings

borealis -is -e northern

bothrio- minutely pitted-

botry- bunched-, panicled-

botryoides, botrys resembling a bunch of grapes

brachiatus -a -um arm-like, branched at about a right angle

brachy- short-

brachybotrys short-clustered

bracteatus -a -um with bracts, bracteate (e.g. inflorescences of *Hydrangea, Poinsettia, Acanthus* etc.)

bracteosus -a -um with large or conspicuous bracts

brasiliensis -is -e from Brazil, Brazilian

brassici- cabbage-

brevi-, brevis -is -e short-

Briza an ancient Greek name for rye, food grain

-bromus -smelling, -stinking

bronchialis -is -e throated, of the lungs (medicinal use)

brumalis -is -e of the winter solstice, winter-flowering

brunneus -a -um russet-brown

bryoides moss-like

Bryonia a name used by Dioscorides

bubalinus -a -um of oxen, of cattle

buccinatorius -a -um, buccinatus -a -um trumpet-shaped, horn-shaped

bucephalus -a -um bull-headed

bucharicus -a -um from Bokhara, Turkistan

bufonius -a -um of toads, living in damp places

bulbifer -era -erum producing bulbs (often when these replace normal flowers)

bulbi-, bulbo- bulb-

Bulbine the Greek name for a bulb

bulbocastanus -a -um chestnut-brown-bulbed

bulbosus -a -um swollen, having bulbs, bulbous

bullatus -a -um with a bumpy surface, puckered, blistered, bullate

bumannus -a -um having large tubercles

-bundus -a -um -doing, -having the capacity for

buphthalmoides ox-eyed

Butomus ox cutter, a name used by Theophrastus with reference to the sharp-edged leaves *butyraceus -a -um* oily, buttery

Butyrospermum butter-seed (the oily seed of the shea butter tree)

buxifolius -a -um box-leaved

Buxus a name used by Virgil

byzantinus -a -um from Istanbul (Byzantium), Turkish

cacao Aztec name for the cocoa tree, *Theobroma*

cachemiricus -a -um from Kashmir

cacti- catus-like- (originally the Greek cactus was an Old World spiny plant, not a cactus)

cacumenus -a -um of the mountain top

cadmicus -a -um with a metallic appearance

caducus -a -um transient, not persistent

caeno-, caenos- fresh-, recent-

caenosus -a -um muddy, growing on mud

caerulescens turning blue, bluish

caeruleus -a -um dark sky-blue

caesius -a -um bluish-grey, lavender-coloured

caespitosus -a -um growing in tufts, matted, tussock-forming

caffer -era -erum, cafforus -a -um of South Africa, of the Kaffirs (unbelievers)

cainito the West Indian name for the star apple

Caiophora burn carrying (the stinging hairs)

cairicus -a -um from Cairo, Egypt

cajan the Malay name for pigeon pea

Cajanus the Malay name

cajennensis -is -e the Cayenne, French Guiana

cajuputi the Malayan name

cala- beautiful-

calaba the West Indian name

calabricus -a -um from Calabria, Italy

calamarius -a -um reed-like, resembling *Calamus*

calaminaris -is -e cadmium-red, growing on the zinc ore calamine

Calamintha beautiful mint

calamitosus -a -um dangerous, causing loss

Calanthe beautiful flower

calathinus -a -um basket-shaped

calcaratus -a -um spurred, having a spur

calcareus -a -um of lime-rich soils

calcatus -a -um with a spur

calceolatus -a -um slipper-shaped, shoe-shaped

calceolus -a -um like a small shoe

calcicolus -a -um living on limy soils

calcifugus -a -um disliking lime

caledonicus -a -um from Scotland (Caledonia), Scottish, of northern Britain

Calendula First day of the month (medieval connection with paying accounts and settling debts)

calidus -a -um fiery, warm

californicus -a -um from California U.S.A.

caliginosus -a -um of misty places

Calla a name used in Pliny, beauty

calli-, callis- beautiful-

Calliandra beautiful stamens (shaving brush tree)

callifolius -a -um *Calla*-leaved

callimorphus -a -um of beautiful form or shape

Callistemon beautiful stamens (bottle brush tree)

Callistephus beautiful crown

callistus -a -um very beautiful

Callitriche beautiful hair

callosus -a -um hardened

Calluna brush, cleanser (former use for sweeping)

calo- beautiful-

calomelanos beautifully dark

calophrys with dark margins

calpophilus -a -um estuary-loving, estuarine

calvescens becoming bald, with non-persistent hair

calvus -a -um hairless, bald, naked

calyc, calyci- calyx-

calycinus -a -um calyx-like, having a persistent calyx

calyculatus -a -um resembling a small calyx

calyptr-, calyptro- hooded-, lidded-

calyptratus -a -um with a cap-like cover over the flower or fruits

camaldolulensis -is -a from the Camaldoli gardens near Naples

camara a West Indian name, arched

cambrensis -is -e, cambricus -a -um from Wales (Cambria), Welsh

camelliiflorus -a -um *Camellia*-flowered

cammarus -a -um from a name used by Dioscorides, lobster

campani-, campanularius -a -um, campanulatus -a -um, campanulus -a -um bell-shaped, bell-flower-like

Campanula little bell

campester -tris -tre, campestris -is -e of the pasture, from flat land

camphoratus -a -um camphor-like scented

campto- bent-

campyl-, campylo- bent-, curved-

camtschatcensis -is -e, camtschaticus -a -um from Kamchatka
   Peninsula, Siberia
canadensis -is -e from Canada, Canadian
canaliculatus -a -um furrowed, channelled
cananga from a Malayan name
canarius -a -um canary-yellow
canariensis -is -e of the Canary Isles, of bird-food
canarinus -a -um yellowish, resembling *Canarium*
cancellatus -a -um cross-banded, chequered
candelabrum -a -um chandelier-like, like a branched candle-stick
candicans whitish, hoary-white
candidus -a -um shining-white
canephorus -a -um like a basket-bearer
canescens turning hoary white, canescent
caninus -a -um of the dog, sharp-toothed or spined, wild or
   inferior
cannabinus -a -um hemp-like, resembling *Cannabis*
cano- hairy-
cantabricus -a -um from Cantabria, N. Spain
cantabriensis -is -e from Cantabrigia or Cambridge
cantianus -a -um from Kent
canus -a -um whitish-grey
capensis -is -e from Cape Colony, South Africa
capillaceus -a -um, capillaris -is -e hair-like, very slender
capillipes with a very slender stalk
capillus-veneris Venus' hair
capitatus -a -um growing in a head, head like (inflorescence)
capnoides smoke-like
cappadocicus -a -um, cappadocius -a -um from Cappadocia, Asia
   Minor
capraeus -a -um, capri- of the goat, goat-like (capraea = she goat)
capreolatus -a -um tendrilled, with tendrils
Capsicum biter (the hot taste)
capsularis -is -e producing capsules
caput-medusae Medusa's head
caracallus -a -um beautiful snail
cardamomum the ancient Greek name for the Indian spice
cardi-, cardio- heart-shaped-
cardiacus -a -um of heart conditions (medicinal use)
cardinalis -is -e cardinal-red
cardunculus -a -um thistle-like
Carex cutter (the sharp leaf margins of many)
caricinus -a -um, caricosus -a -um resembling *Carex*

caricus -a -um from Caria, province of Asia Minor
carinatus -a -um keeled, having a keel-like ridge
carmineus -a -um carmine
carneus -a -um, carnicolor flesh-coloured
carniolicus -a -um from Carniola, Yugoslavia
carnosus -a -um fleshy, thick and soft-textured
carota the old name for carrot (*Daucus carota*)
carpathicus -a -um, carpaticus -a -um from the Carpathian
    Mountains
carpetanus -a -um from the Toledo area of Spain
carpo-, carpos-, -carpus -a -um -fruited, -fruit
carthusianorus -a -um of the Grande Chartreuse Monastery of
    Carthusian Monks, Grenoble, France
carunculatus -a -um with a prominent caruncle (the outgrowth
    of the seed coat, usually obscuring the micropyle)
carvi from Caria, Asia Minor
carvi- carrot-
Carya the ancient Greek name for a walnut
caryo- nut-, clove-
caryophylleus -a -um resembling a stichwort, clove-coloured
caspicus -a -um of the Caspian area
cassioides resembling *Cassia*
cassubicus -a -um from Cassubia, part of Pomerania
castaneus -a -um, castanus -a -um chestnut-brown
castello-piavae for Baron Castello de Piava
castus -a -um spotless, pure
cat-, cata-, cato- below-, downwards-, outwards-, from under-,
    against-, along-
Catabrosa eaten (the appearance of the tips of the lemmas)
catafractus -a -um, cataphractus -a -um enclosed, armoured, closed
    in
catalpa an East Indian name
catarius -a -um of cats (catnip of catmint)
catarractae, catarractarum growing near waterfalls, resembling a
    waterfall
catenarius -a -um, catenatus -a -um chain-like
catharticus -a -um purgative, purging, cathartic
cathayanus -a -um, cathayensis -is -e from China
catholicus -a -um world-wide, universal, of Catholic lands (Spain
    and Portugal)
caucasiacus -a -um from the Caucasus
caudatus -a -um, caudi- tailed (see Fig. 7(a))
caudiculatus -a -um with a thread-like caudicle or tail

65

caulescens having a distinct stem

cauliatus -a -um, -caulis -is -e, -caulo, -caulos of the stem or stalk, -stemmed, -stalked

cauliflorus -a -um bearing flowers on the main stem, flowering on the old woody stem

causticus -a -um with a caustic taste (burning the mouth)

cauticolus -a -um growing on cliffs, cliff-dwelling

cavernicolus -a -um growing in caves, cave-dwelling

cavernosus -a -um full of holes

cavus -a -um hollow

celebicus -a -um from the Indonesian Island of Celebes

celeratus -a -um hastened

-cellus -a -um -lesser, -somewhat

Celosia burning (flower colour)

cembra the old name for the arolla or stone pine

cembroides, cembrus -a -um resembling the arolla pine (*Pinus cembra*)

cenisius -a -um from Mt Cenis on the French-Italian border

ceno-, cenose- empty-

Centaurea Centaur (mythical creature half man and half bull)

centaurioides resembling *Centaurea*

Centaurium for the Centaur, Chiron, who was fabled to have used this plant medicinally

centi- one hundred-, many-

centra-, centro-, -centrus -a -um -spurred

centralis -is -e in the middle, central

Centranthus spur flower

cepa the old name for an onion

cepaeus -a -um growing in gardens, from the ancient Greek for a salad plant

cephal- head-, head-like-

cephalidus -a -um having a head

cephalonicus -a -um from Caphalonia, one of the Ionian Islands

cephalotes having a small head-like appearance

cephalotus -a -um with flowers in a large head

-cephalus -a -um -headed

-ceras -horned

ceraseus -a -um waxy

cerasifer bearing cherries (or cherry-like fruits)

cerasinus -a -um cherry-red

Cerastium horned (the shape of the fruiting capsule)

Cerasus from an Asiatic name for the sour cherry

cerato- horn-shaped-

Ceratophyllum horned leaf
Ceratopteris horned fern
cerealis -is -e for Ceres, the goddess of agriculture
cereus -a -um waxy (*cereus* = wax taper)
cerifer -era -erum wax-bearing
cerinus -a -um waxy
cernuus -a -um drooping, curving forwards
Ceropegia fountain of wax (appearance of inflorescence)
cerris the ancient Latin name for the turkey oak
cervicarius -a -um constricted, keeled
cervinus -a -um tawny, stag-coloured
cespitosus -a -um see *caespitosus -a -um*
Cestrum from an ancient Greek name
Ceterach an Arabic name for a fern
cevisius -a -um closely resembling
chaeno- splitting-, gaping-
Chaenomeles gaping apple
chaero- pleasing-
chaeto- long hair-like-
chalcedonicus -a -um from Chalcedonia, Turkish Bosphorus
chamae- on the ground-, ground hugging-, prostrate-, low-
   growing-, lowly-
Chamaecyparis dwarf cypress
chamaedrys ground oak
Chamaenerion dwarf oleander
chamaeunus -a -um lying on the ground
characias the name in Pliny for a spurge with very caustic latex
charantius -a -um graceful
charianthus -a -um with elegant flowers
chartaceus -a -um parchment-like
chasmanthus -a -um having open flowers
chasmophilus -a -um liking hollows or chasms
chauno- gaping-
cheilanthus -a -um lipped-flower
cheilo- lip-
cheiri- red-flowered (from an Arabic name)
cheiro- hand-
Chelidonium swallow (flowers at the season when swallows
   arrive)
Chenopodium goose-foot (the shape of the leaves)
cherimolia a Peruvian-Spanish name
chia from the Greek island of Chios
chilensis -is -e from Chile, Chilean

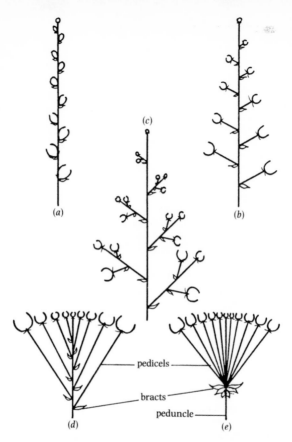

Fig. 2. Types of inflorescence which provide specific epithets:
(a) a spike (e.g. *Actaea spicata* L. and *Phyteuma spicatum* L.); (b) a
raceme (e.g. *Bromus racemosus* L. and *Sambucus racemosa* L.); (c) a
panicle (e.g. *Carex paniculata* L. and *Centaurea paniculata* L.); (d) a
corymb (e.g. *Silene corymbifera* Bertol. and *Teucrium corymbosum*
R.Br.); (e) an umbel (e.g. *Holosteum umbellatum* L. and *Butomus
umbellatus* L.).
In these inflorescences the oldest flowers are towards the periphery
or the base.

chiloensis -is -e from Chiloe Island off Chile
-chilus -a -um -lipped
chima-, chimon- winter-
chimaera monstrous, fanciful
Chimonanthus winter flower
chinensis -is -e from China, Chinese
chio-, chion-, chiono- snow-
chionanthus -a -um snow-white-flowered
Chionedoxa glory of the snow (very early flowering)
chironius -a -um after Chiron, the centaur of Greek mythology
   who taught the medicinal use of plants
chirophyllus -a -um with hand-shaped leaves
clamy- cloak-
chlor-, chloro-, chlorus -a -um yellowish-green
chloranthus -a -um green-flowered
Chloris for Chloris, the Greek goddess of flowers
chlorophyllus -a -um green-leaved
chocolatinus -a -um chocolate-brown
chondro- rough-, granular-, lumpy-, coarse-
chordatus -a -um cord-like
chordo- string-, slender-elongate-
chori- separate-, apart-
-chromatus -a -um, -chromus -a -um -coloured
chrono- time-
chrysanthus -a -um yellow-flowered
chryseus -a -um, chrys-, chryso- golden-yellow
chrysographes marked with gold lines, as if written upon in gold
chrysops with a golden eye
chrysostomus -a -um with a golden throat
chyllus -a -um from a Himalayan vernacular name
chylo- sappy-
cibarius -a -um edible
cicer the old Latin name for the chick-pea
Cicerbita an old Italian name for a thistle
Cichorium from an Arabic name
ciconius -a -um resembling a stork's neck
cicutarius -a -um resembling *Cicuta*, with large two- to three-
   pinnate leaves
Cicuta the Latin name for hemlock
ciliaris -is -e, ciliatus -a -um, ciliosus -a -um fringed with hairs
cilicicus -a -um from Cilicia in southern Turkey
-cillus -a -um -lesser
cincinnatus -a -um with crisped hairs

cinctus -a -um girdled

cineraceus -a -um, cinerarius -a -um, cinerascens ash-coloured, covered with ash-grey felted hairs

cinereus -a -um ash-grey

cinnabarinus -a -um cinnabar-red

cinnamomeus -a -um cinnamon-brown

circinalis -is -e, circinatus -a -um curled-round, coiled crozier-shaped

circum- around-

cirratus -a -um, cirrhatus -a -um, cirrhiferus -a -um having or carrying tendrils

cirrhosus -a -um with large tendrils

Cirsium the ancient Greek name for a thistle

Cissus the ancient Greek name for ivy

citreus -a -um, citrinus -a -um citron-yellow

citri- citron-like-

citriodorus -a -um citron-scented

citrodorus -a -um lemon-scented

Cladium small branch

clado- shoot-, branch-, of the shoots-

clandestinus -a -um concealed, hidden, secret

clausus -a -um shut, closed

clavatus -a -um, clavi-, clavus -a -um club-shaped, clubbed

claviculatus -a -um having tendrils, tendrilled

clavigerus -a -um club-bearing

cleio-, cleisto- shut-, closed-

Clematis the Greek name for several climbing plants

clematitis -is -e vine-like, with long vine-like twiggy branches

Clianthus glory flower (the glory pea)

clino- prostrate-, bed-

clipeatus -a -um shield-shaped

clivorum of the hills

clymenus -a -um from an ancient Greek name

clypeatus -a -um, clypeolus -a -um like a Roman shield

cneorus -a -um of garlands, the Greek name for an olive-like shrub

Cnicus the Greek name for a thistle used in dyeing

co-, col-, con- together-, together with-, firmly-

coaetaneus -a -um ageing together (leaves and flowers both senesce together)

coagulans curdling

coarctatus -a -um pressed together, bunched, contracted

coca the name used by South American Indians

cocciferus -a -um bearing berries

coccineus -a -um scarlet (the dye produced from galls on *Quercus coccifera*)

-coccus -a -um -berried

Cochlearia spoon (shape of the basal leaves)

cochlearis -is -e spoon-shaped

cochleatus -a -um twisted like a snail shell, cochleate

cochlio-, cochlo- spiral-, twisted-

-codon -bell, -mouth

Codonopsis bell-like (flower shape)

coelestinus -a -um, coelestis -is -e, coelestus -a -um, coeli- sky-blue, heavenly

coeli-rosa rose of heaven

coelo- hollow-

Coeloglossum hollow-tongue (the lip of the flower)

coen-, coenos- common-

coerulescens bluish

coeruleus -a -um blue

coggygria the ancient Greek name for *Cotinus*

cognatus -a -um closely related to

Coix the ancient Greek name for Job's tears grass

-cola -loving, -inhabitant of (preceded by a habitat)

colchicus -a -um from Colchis in the Caucasus

coleatus -a -um sheath-like

coleo- sheath-

Coleus sheath (the filaments around the style)

collinus -a -um from the hills, growing on hills

-collis -is -e- necked

colocynthis -is -e ancient Greek name for *Citrullus colocynthis*

colombinus -a -um dove-like

colonus -a -um forming a mound, humped

colorans, coloratus -a -um coloured

colubrinus -a -um snake-like

columbarius -a -um, columbinus -a -um dove-like, dove-coloured, of doves

Columnea for Fabio Colonna of Naples who published *Phyto-basanos* in 1592

columnaris -is -e columnar, pillar-like

-colus -a -um -dwelling (follows a kind of habitat or place)

Colutea a name used by Theophrastus for a tree

com- with-, together with-

comans, comatus -a -um hairy-tufted

commixtus -a -um mixed together, mixed up

communis -is -e common, growing in company, clumped

commutatus -a -um changed, altered (e.g. from previous inclusion in another species)

comorensis -is -e from Comoro Island, East Africa

comosus -a -um shaggy-tufted, with tufts formed from hairs or leaves or flowers

compactus -a -um close-growing, closely packed together

complanatus -a -um flattened out upon the ground

compositus -a -um with flowers in a head, *Aster*-flowered, compound

compressus -a -um flattened sideways (as in stems), pressed together

comptus -a -um ornamented, with a headdress

con- with-, together with-

concatenans, concatenatus -a -um joined together, forming a chain

concavus -a -um basin-shaped, concave

concinnus -a -um well-proportioned, neat, elegant

concolor uniformly coloured

condensatus -a -um crowded together

condylodes knobbly, with knuckle-like bumps

confertus -a -um crowded, pressed together

conformis -is -e symmetrical, conforming to type or relationship

confusus -a -um easily mistaken for another species, intricate

congestus -a -um arranged very close together

conglomeratus -a -um clustered, crowded together

conicus -a -um cone-shaped, conical

conifer -era -erum cone-bearing

conii- hemlock-like, resembling *Conium*

Conium the Greek name for hemlock plant and poison

conjugalis -is -e, conjugatus -a -um joined together in pairs, conjugate

conjunctus -a -um joined together

connatus -a -um united at the base

connivens converging, connivent

cono- cone-shaped-

conoideus -a -um cone-like

Conopodium cone foot

conopseus -a um cloudy, gnat-like

consanguineus -a -um closely related, of the same blood

consimilis -is -e much resembling

Consolida make firm, the ancient Latin name from its use in healing medicines

consolidus -a -um stable, firm

conspersus -a -um speckled, scattered
conspicuus -a -um easily seen, conspicuous
constrictus -a -um erect, dense
contemptus -a -um despising, despised
contiguus -a -um close and touching, closely related
contortus -a -um twisted, bent
contra-, contro- against-
contractus -a -um drawn together
controversus -a -um doubtful, controversial
Convallaria valley (from the natural habitat of lily-of-the-valley)
convalliodorus -a -um Lily-of-the-valley scented
conversus -a -um turning towards, turned together
convexus -a -um humped, bulged outwards
convolutus -a -um rolled together
convolvulus entwine, twined together
Conyza a named used by Theophrastus
copallinus -a -um from a Mexican name, yielding copal gum
copiosus -a -um abundant, copious
copticus -a -um from Egypt, Egyptian
coracinus -a -um raven-black
corallinus -a -um, corallioides coral-red
corbularia like a small basket
Corchorus the Greek name for jute
cordatus -a -um, cordi- heart-shaped, cordate (see Fig. 6(*e*))
cordifolius -a -um with heart-shaped leaves
Cordyline club (some have large club-shaped roots)
coreanus -a -um from Korea, Korean
Coreopsis bug-like (the shape of the fruits)
coriaceus -a -um tough, leathery, thick-leaved
Coriandrum the Greek name for a flavouring plant
Coriaria leather (used in tanning)
coriarius -a -um of tanning, leather-like
corrid- Coris-like-
corii- leathery-
coritanus -a -um resembling *Coris*, from the East Midlands (home
    area of the Coritani tribe of ancient Britons)
corneus -a -um horny
corni-, cornifer -era -erum, corniger -era -erum horn-bearing,
    horned
corniculatus -a -um having horn- or spur-like appendages or
    structures
cornubiensis -is -e from Cornwall, Cornish
cornucopius -a -um shaped like the horn of plenty

Cornus the Latin name for the cornelian cherry

cornutus -a -um horned, horn-shaped

corollinus -a -um with a conspicuous corolla

coronarious -a -um forming a crown, garlanding

coronatus -a -um crowned

Coronilla little crown

Coronopus Theophrastus' name for crowfoot (leaf shape)

Corrigiola shoe thong (the slender stems)

corsicus -a -um from Corsica, Corsican

Cortaderia cutter (from the Spanish-American name which
  refers to the sharp-edged leaves)

corticalis -is -e, corticosus -a -um with a notable or pronounced
  bark

coryandrus -a -um with helmet-shaped stamens

Corydalis crested lark (the spur of the flowers)

corylinus -a -um, coryli-, Corylopsis hazel-like, resembling *Corylus*

Corylus the Latin name for the hazel

corymbosus -a -um with flowers arranged in a corymb, with a
  flat-topped raceme (see Fig. 2(*d*))

coryne-, coryno- club-, club-like-

corynephorus -a -um bearing a club, clubbed

coryph- at the summit-

corys-, -corythis -helmet, -cuculate

Cosmos beautiful

costalis -is -e with prominent ribs

Cotoneaster quince-like

Cotula small cup

Cotyledon cup-shaped, hollow (the leaf structure)

coum from a Hebrew name

cracca ancient Greek name for a vetch, also used in Pliny

Crambe ancient Greek name for a cabbage-like plant

crassi— thick-, fleshy-

Crassicaulis -is -e thick-stemmed

Crassula somewhat thick

crassus -a -um, thick, fleshy

Crataegus, crateri, cratero- strong, a name used by Theophrastus

creber -ra -rum, crebri- densely clustered, frequently

crenati-, crenatus -a -um with small rounded teeth (the leaf
  margins, see Fig. 4(*a*))

crepidatus -a -um sandal- or slipper-shaped

crepitans rustling

cretaceus -a -um of chalk, inhabiting chalky soils

creticus -a -um from Crete, Cretan

criniger -era -erum carrying hairs
crinitus -a -um with a tuft of long soft hairs
crispatus -a -um closely waved, curled
crispus -a -um with a waved or curled margin
crista-galli cock's comb (the crested bracts)
cristatus -a -um tassel-like at the tips, crested
Crithmum barley (similarity of the seed)
crocatus -a -um citron-yellow, saffron-like (used in dyeing)
croceus -a -um saffron-coloured, yellow
Crocosmia saffron-scented (the dry flowers)
Crocus from the Chaldean name
Crotalaria rattle (seeds loose in the inflated pods of some)
cruciatus -a -um arranged cross-wise (the leaf arrangement)
crucifer -era -erum in the form of a cross, cruciform
cruentus -a -um blood-coloured
crumenatus -a -um pouched
crurus -a -um leg-shaped
crus-andrae St Andrew's cross
crus-galli cock's spur or leg
crustatus -a -um encrusted
crus-maltae Maltese cross
crypto- hidden-, obscurely-
Cryptogramma hidden writing (the concealed sori)
crystallinus -a -um with a glistening surface, as though covered
    with crystals
cteno- comb-
cubitalis -is -e as long as the forearm plus the hand, a cubit tall
cucubalus a name used by Dioscorides and Pliny
cucullaris -is -e, *cucullatus -a -um* hooded, hood-like
cucumerinus -a -um resembling cucumber
cucurbitinus -a -um melon- or marrow-like
cujete a Brazilian name
cultratus -a -um, *cultriformis -is -e* shaped like a knife-blade
cultus -a -um cultivated, grown
-culus -a -um -lesser
-cundus -a -um -able, -dependable
cuneatus -a -um, *cuneiformis -is -e* wedge-shaped, narrow below
    and wide above
Cuphea curve (the fruiting capsule)
cupreatus -a -um coppery, bronzed
cupressinus -a-um cypress-like
Cupressus symmetry (the conical shape)
cupreus -a -um copper-coloured, coppery

cupularis -is -e cup-shaped

curassavicus -a -um from Curacao, West Indies

curcas the ancient Latin name for *Jatropha*

Curculigo weevil (the beak of the fruit)

Curcuma the Arabic name for turmeric

curti-, curtus -a -um short, broken short

curvatus -a -um curved

Cuscuta the medieval name for dodder

cuspidatus -a -um, cuspidi- abruptly narrowed into a short rigid point

cyaneus -a -um, cyano-, cyanus Prussian-blue

Cyanotis blue ear

Cyclamen circled (the twisted fruiting stalk)

cyclamineus -a -um resembling *Cyclamen*

cycl-, cyclo- circle-, circular-

cyclius -a -um round, circular

cyclops gigantic

Cydonia the Latin name of a tree from Cydon, Crete

cylindricus -a -um, cylindro- long and round, cylindrical

cymbalarius -a -um cymbal-like (the leaves of Toad flax)

cymbidi- boat-

cymbiformis -is -e boat-shaped

cymosus -a -um having the flowers borne in a cyme (see Fig. 3)

cynanchicus -a -um of quinsy (literally dog-throttling) from its former medicinal use

Cynapium dog-celery, dog parsley

cyno- dog- (usually has derogatory undertone)

Cynodon dog-tooth

cynops the ancient Greek name for a plantain

Cynosurus dog-tail

cyparissias cypress-leaved, used in Pliny for spurge

Cypripedium Aphrodite's foot (slipper)

cyrt- curved, arched-

Cyrtomium bulged (the leaflets)

cyst-, cysti-, cysto- hollow-, pouched-

Cystopteris bladder fern

dactyl-, dactylo-, dactyloides finger-, finger-like-

Dactylorchis finger orchid (the arrangement of the root-tubers)

dahuricus -a -um, dauricus -a um, davuricus -a -um from Dauria, N. E. Asia

dalmaticus -a -um from Dalmatia, eastern Adriatic, Dalmatian

damascenus -a -um from Damascus, coloured like *Rosa damascena damasonium* a name in Pliny for *Alisma*

Danae after the daughter of Acrisius Persius, in Greek mythology

danfordiae for Mrs C. G. Danford

danicus -a -um from Denmark, Danish

Daphne old name for bay laurel, from that of a nymph, in Greek mythology

dasy- thick-, thickly hairy-

dasyclados shaggy-twigged

dasyphyllus -a -um thick-leaved

-dasys -hairy

Datura from an Indian vernacular name

dauci- carrot-like-

Daucus the Latin name for the carrot

davidii, davidianus -a -um for l'Abbé Armand David

de- downwards-, outwards-, from-

dealbatus -a -um with a white powdery covering

debilis -is -e weak, feeble

dec-, deca- ten-

decandrus -a -um ten-stamened

deciduus -a -um not persisting beyond one season

decipiens deceptive, deceiving

declinatus -a -um curved downwards, turned aside

decolorans staining, discolouring

decompositus -a -um divided more than once (leaf structure)

decoratus -a -um, decorus -a -um handsome, decorative

decorticans, decorticus -a -um with shedding bark

decumanus -a -um very large (refers literally to one tenth of a division of Roman soldiers)

decumbens prostrate with turned up tips

decurrens running down (e.g. the bases of leaves down the stem), decurrent

decussatus -a -um at right angles (as when leaves are in two alternating ranks)

deflexus -a -um bent sharply downwards

deformis -is -e misshapen

dehiscens splitting open, gaping, dehiscent

dejectus -a -um debased

delectus -a -um chosen, choice

delicatissimus -a -um most charming, most delicate

deliciosus -a -um of pleasant flavour

Delonix obvious claw (on the petals)

delphicus -a -um from Delphi, Greece, Delphic

Delphinium dolphin head (a name used by Dioscorides)

deltoides, deltoideus -a -um triangular-shaped, deltoid

demersus -a -um underwater, submerged

demissus -a -um hanging down, low

dendri-, dentro- tree-, tree-like-, on trees-

dendricolus -a -um tree-dwelling

dendroideus -a -um, dendromorphus -a -um tree-like

densatus -a -um, densi-, densus -a -um crowded, dense, close (habit of stem growth)

dens-canis dog's tooth

dens-leonis lion's tooth

Dentaria toothwort (the signature of the scales upon the roots)

dentatus -a -um, dentifer -era -erum, dentosus -a -um having teeth, with outward-pointing teeth, dentate (see fig. 4(*b*))

denudatus -a -um hairy or downy but becoming naked, denuded

deodarus -a -um from the Indian State of Deodar

deorsus -a -um downwards, hanging

deorum of the gods

depauperatus -a -um imperfectly formed, dwarfed, of poor appearance

dependens hanging down

depressus -a -um flattened downwards

derelictus -a -um abandoned, neglected

deserti, desertorum of deserts, inhabiting deserts

desma- bundle-

Desmanthus bundle flower (appearance of the inflorescence)

detergens delaying

detersus -a -um wiped clean

detonsus -a -um shaved, bald

deustus -a -um burned

di-, dis- two-, twice-, between-, away from-

dia- through-, across-

diabolicus -a -um devilish, slanderous, two horned

diacritus -a -um distinguished, separated

diademus -a -um band or fillet, crown

dialy- very deeply incised to separate

diandrus -a -um two-stamened

Dianthus Jove's flower (a name used by Theophrastus)

Diapensia formerly a name for sanicle but re-applied by Linneaus

diaphanoides resembling (leaves of) *Hieracium diaphanum*

diaphanus -a -um transparent (leaves)

Dicentra twice-spurred

dicha-, dicho- double-, into two-

Dichorisandra two separate men (two stamens diverge from the remainder)

dichotomus -a -um repeatedly divided into two equal portions

dichrano- two-branched

dichranotrichus -a -um with two-pointed hairs

dichroanthus -a -um with two-coloured flowers

dichromus -a -um, dichrous -a -um of two colours

dicoccus -a -um having two nuts or two-berried

dictyo- netted-

didymo-, didymus -a -um twin-, twinned-, double-, equally divided-

dielsianus -a -um, dielseii for F.L.E. Diels of the Berlin Botanic Garden

Dierama funnel (the shape of the perianth)

difformis -is -e of unusual form or shape

diffusus -a -um loosely spreading, diffuse

Digitalis from the German name Fingerhut (meaning a thimble)

digitatus -a -um fingered, lobed from a single point, digitate

digraphis -is -e twice-inscribed, lined with two colours

dilatatus -a -um, dilatus -a -um widened, spread out, dilated

dimidiatus -a -um with two equal parts, dimidiate

dimorpho-, dimorphus -a -um two-shaped, with two forms (of leaves or flowers or fruits)

diodon two-toothed

dioicus -a -um of two houses, having separate male and female plants, dioecious

Dioscorea for Dioscorides (type genus of the yam family)

Diotis two-ears (the spurs of the corolla)

diphyllus -a -um two-leaved

Diplotaxis two-places (the two-ranked seeds)

diplotrichus -a -um, diplothrix having two kinds of hairs

Dipsacus dropsy (the water which collects in the leaf bases)

diptero-, dipterus -a -um two-winged

disci-, disco- disk-

discerptus -a -um torn apart (the leaves)

discoideus -a -um disk-like, discoid

discolor of different colours

dispar unequal, different

dispersus -a -um scattered

dissectus -a -um cut into many deep lobes

dissimilis -is -e unlike

dissitiflorus -a -um in loose heads, not compact (the flowers)

distachyon, distachyus -a -um two-branched, two-spiked, with two spikes

distans widely separated

distichus -a -um two-ranked, arranged in two rows

distillatorius -a -um shedding drops, of the distillers

distortus -a -um malformed, grotesque

Distylium two styles (the separate styles)

distylus -a -um two-styled

diurnus -a -um lasting for one day, day-flowering

diutinus -a -um long-lasting

divaricatus -a -um wide spreading, straggling

divensis -is -e of Chester (Deva)

divergens spreading out

diversi- differing-, diversely-

divionensis -is -e from Dijon, France

divisus -a -um divided

divulsus -a -um torn violently apart

divus -a -um belonging to the gods

dodec- twelve-

dolabratus -a -um, dolabriformis -is -e hatchet-shaped

dolicho- long-

Dolichos the ancient Greek name for long-podded beans

dolichostachyus -a -um long-spiked

dolosus -a -um deceitful

domesticus -a -um of the household

donax an old Greek name for a reed

Doronicum from an Arabic name

-dorus -a -um -bag, -bag-shaped

dory- spear-

Dorycnium ancient Greek for a *Convolvulus* re-applied by Dioscorides

-doxa -glory

Draba a name used by Dioscorides

Dracaena female dragon

Dracunculus little dragon (a name used by Pliny)

drepanus -a -um, drepano- sickle-shaped

Drimia acrid

Drimys acrid (taste of the bark)

Drosera dew (the glistening glandular hairs)

drupaceus -a -um stone-fruited

Dryas oak-nymph (leaf shape)

drymo- wood-, woody-

dryophyllus -a -um oak-leaved
Dryopteris oak fern (the habitat)
dubius -a -um uncertain, doubtful
dulcamarus -a -um bitter-sweet
dulcis, -is -e sweet-tasted, mild
dumalis -is -e, dumosus -a -um compact, thorny, bushy
dumetorum of bushy habitats, of thickets
dumnoniensis -is -e from Devon, Devonian
dunensis -is -e of sand dunes
duplex, duplicatus -a -um double, duplicate
duracinus -a -um hard-fruited
duriusculus -a -um rather hard
durus -a -um hard
dys- poor-, ill-, bad-, difficult-
Dyschoriste poorly divided (the stigma)
dysentericus -a -um of dysentery (medicinal treatment for)
dyso- evil-smelling
Dysodea evil smelling

e-, ef- ex- without-, not-, from out of-
ebenaceus -a -um ebony-like
ebenus -a -um ebony-black
eboracensis -is -e from York (Eboracum)
ebracteatus -a -um without bracts
Ebulus a name in Pliny for danewort
eburneus -a -um ivory-white
ecae for Mrs E. C. Aitchison
Ecballium expeller (the discharge of the squirting cucumber)
Eccremocarpus hanging fruit
echinatus -a -um, echino- covered with prickles, hedgehog-like
Echinochloa hedgehog grass (the awns)
echioides resembling *Echium*
Echium a name used by Dioscorides
eclecteus -a -um selected, picked out
ect-, ecto- on the outside, outwards-
edentatus -a -um without teeth, toothless
edulis -is -e of food, edible
effusus -a -um spread out
eglantarius -a -um from a French name (eglantois)
Elaeagnus olive chaste tree
elasticus -a -um yielding an elastic substance, elastic
elatarius -a -um driving away (squirting out the seeds)
Elatine a name used by Dioscorides (little fir trees)

elatior taller

elatus -a -um tall, high

elegans, elegantulus -a -um graceful

eleo- marsh- (cf. heleo-)

elephantipes like an elephant's foot

eleuthero- free-

ellipsoidalis -is -e ellipsoidal (a solid of oval profile)

ellipticus -a -um about twice as long as broad, elliptic

-ellus -ella -ellum -lesser (diminutive ending)

Elodea marsh (the habitat)

elodes as *helodes*, of bogs and marshes

elongatus -a -um lengthened out

Elymus Hippocrates' name for a millet-like grass

em-, en- in-, into-, for-, within-, not-

emarginatus -a -um notched at the apex (see Fig. 7(*h*))

emasculus -a -um without functional stamens

Embothrium in little pits (the anthers)

emerus -a -um from an early Italian name for a vetch

emeticus -a -um causing vomiting, emetic

eminens noteworthy, outstanding, prominent

emodensis -is -e, emodi- from Mount Emodus, N. India

Empetrum in rocks (Dioscorides' name refers to the habitat)

Enarthrocarpus pointed fruit

encephalo- in a head-

endivia ancient Latin name for chicory (see *Intybus*)

Endymion for Diana's lover, of Greek mythology

emodensis -is -e, emodi from Mount Emodus, N. India

ennea- nine-

ensatus -a -um, ensi- sword-shaped

-ensis -is-e -from, -of (a place)

ento-, endo- on the inside-, inwards-

entomophilus -a -um insect-loving, of insects

ep-, epi- upon-, on-, over-, somewhat-

epeteius -a -um annual

Epidendrum tree-dweller (the epiphytic habit)

epigaeus -a -um growing close to the surface of the ground

epigeios of dry earth, from dry habitats

epihydrus -a -um of the water surface

Epilobium Gesner's name indicating the positioning of the corolla
    on top of the ovary

Epimedium the name used by Dioscorides

epiphyticus -a -um growing upon another plant

epipsilus -a -um somewhat naked (the sparse foliage)

epithymum parasitic upon thyme
equalis -is -e equal
equestris -is -e of horses or horsemen, equestrian
equinus -a -um of horses
Equisetum horse hair, a name used in Pliny
equitans astride as on horseback (e.g. the leaves of iris)
Eragrostis love grass
Eranthis spring flower
erectus -a -um upright, erect
eremi-, eremo- solitary-, deserted-
Eremurus solitary tail
eri-, erio- woolly-
Erica the Latin name
ericetorum of heathland
ericinus -a -um, ericoides heath-like, resembling *Erica*
erigenus -a -um Irish-born
Erigeron early old man (Theophrastus' name)
Erinus Dioscorides' name for a basil-like plant
Eriocaulon woolly stem
Eriophorum wool carrier (cotton grass)
eriophorus -a -um bearing wool
eristhales very luxuriant, *Erithalis*-like
ermineus -a -um ermine-coloured
Erodium heron (the shape of the fruit)
Erophila spring lover
erosus -a -um jagged, as if nibbled irregularly
erraticus -a -um differing from the type, of no fixed habitat
erromenus -a -um vigorous
erubescens blushing, turning red
Eruca belch (the ancient Latin name)
erucastrum colewort-flowered
Ervum the Latin name for orobus, a vetch
Eryngium Theophrastus' name for a spine-leaved plant
Erysimum a name used by Theophrastus
erythraeus -a -um, erythro- red-
Erythroxylon red wood
-escens -ish, -becoming
esculentus -a -um tasty, good to eat, edible
estriatus -a -um without stripes
esula an old generic name from Rufinus
-etorus -a -um -community (indicating the habitat)
etruscus -a -um from Tuscany (Etruria), Italy
ettae for Miss Etta Stainbank

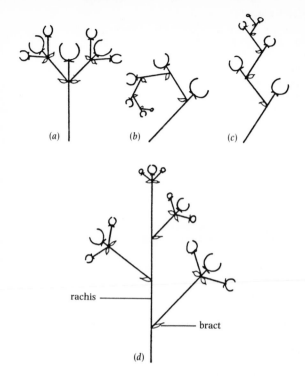

Fig. 3. Types of inflorescence which provide specific epithets: (a), (b) and (c) are cymes with the oldest flower in the centre or at the tip of the inflorescence (e.g. *Saxifraga cymosa* Waldst. & Kit.); (b) may have a three-dimensional form of a screw, or *bostryx*; (c) may be coiled, or *scorpioid* (e.g. *Myositis scorpioides* L.); (d) a raceme of cymes, or thyrse (e.g. *Ceanothus thyrsiflorus* Eschw.).

eu- good-, proper-, completely-, well marked-

euboeus -a -um from the Greek island of Euboea

Eucharis full of grace

euchlorus -a um of a beautiful green, true green

euchromus -a -um well-coloured

Euclidium well-closed (the fruit)

Eucomis beautiful head

eudoxus -a -um of good character

Eulophia beautifully crested

Euonymus famed (Theophrastus' name which also appears as
    Evonymus)

Eupatorium for Eupator who was King of Pontus

euphlebius -a -um well veined

Euphorbia for Euphorbus who used the latex for medicinal
    purposes

Euphrasia good cheer (signature of eyebright flowers as of use in
    eye lotions)

eupheus well-grown

euphodus -a -um long-stalked

euprepes good-looking

eupristus -a um comely

eur-, *eury-* broad-, wide-

europaeus -a -um European

-eus -ea -eum -resembling, -belonging to, -noted for

eustachyus -a -um having long trusses of flowers

evanescens quickly disappearing, evanescent

evectus -a -um lifted up, springing out

evertus -a -um overturned

ex- without-, outside-

Exacum a name in Pliny thought to be derived from an earlier
    Gallic word 'exacon'

exaltatus -a -um lofty, very tall

exasperatus -a -um rough, roughened

excavatus -a -um hollowed out

excellens distinguished, excellent

excelsior, excelsus -a -um very tall

excisus -a -um cut out

excorticatus -a -um without bark, stripped

excurrens with a vein extended into a marginal tooth (as on
    leaves)

exiguus -a -um very small, meagre, poor, petty

exilis, -is -e thin, slender, small

eximius -a -um excellent in size or beauty, choice

85

exitiosus -a -um fatal, deadly, pernicious, destructive

exo- outside-, beyond-, over and above-

Exochorda outside cord (the vascular anatomy of the wall of the ovary)

exoniensis -is -e from Exeter, Devon

exotericus -a -um external, common

exoticus -a -um from a foreign land, not native

expansus -a -um spread out, expanded

explodens exploding

exscapus -a -um without a stem

exsculptus -a -um dug out

exsertus -a -um projecting, protruding

extensus -a -um wide, extended

extra- outside-, beyond-, over and above-

exudans exuding

faba the old Latin name for the broad bean

fabaceus -a -um, fabarius -a -um bean-like

facetus -a -um elegant

faenus -a -um of hay

fagi- beech-like-

Fagopyrum beech wheat (from the Dutch Boekweit)

Fagus the Latin name of the beech

Falcaria sickle (the shape of the leaf segments)

falcatus -a -um, falci-, falciformis -is -e sickle-shaped, falcate

fallax false, deceitful, deceptive

farcatus -a -um solid, not hollow

farfara an old generic name for butterbur

fargesii for Paul Guillaume Farges

farinaceus -a -um with a mealy texture, yielding farina (starch)

farinosus -a -um with a mealy surface

farleyensis -is -e from Farley Hill Gardens, Barbados, West Indies

farnesianus -a -um from the Farnese Palace gardens of Rome

fasciatus -a -um bound together, bundled, fasciated (as in the inflorescence of *Celosia argentea*)

fascicularis -is -e, fasciculatus -a -um clustered in bundles

fastigiatus -a -um with erect branches

fastuosus -a -um proud

Fatsia a Japanese name

fatuus -a -um not good, foolish, insipid

febrifugus -a -um fever-dispelling (medicinal property)

fecundus -a -um fruitful, fecund

felix fern

felix-femina female fern

felix-mas male fern

felosmus -a -um foul-smelling

fenestralis -is -e, fenestratus -a -um with window-like holes or
openings

fennicus -a -um from Finland (Fennica), Finnish

-fer, -ferus -a -um -bearing

ferax fruitful

ferox very prickly

ferreus -a -um iron-hard, durable

ferrugineus -a -um rusty-brown in colour

fertilis -is -e fruitful, heavy-seeding, fertile

Ferula rod (the classical Latin name)

ferulaceus -a -um fennel-like

ferus -a -um wild

festalis -is -e, festinus a -um, festivus -a -um agreeable, bright,
pleasant, cheerful, festive

Festuca straw (a name used in Pliny)

festus -a -um sacred, used for festivals

fetidus -a -um stinking, foetid

fibrillosus -a -um, fibrosus, -a -um with copious fibres, fibrous

ficaria an old generic name for the lesser celandine (small
figs – the root tubers)

fici-, ficoides fig-like

ficto-, fictus -a -um false

Ficus the ancient Latin name for the fig

-fid, -fidus -a -um -cleft, -divided

Filago thread (the medieval name refers to the woolly indumentum)

filamentosus -a -um, filarius -a -um, fili- thread-like, with filaments or threads

filicaulis -is -e having very slender stems

filicinus -a -um, filici-, filicoides fern-like

filiculoides like a small fern

filiformis -is -e thread-like

Filipendula thread-suspended (slender attachment of the tubers)

filix fern

fimbriatus -a -um, fimbri- fringed

firmus -a -um strong, firm

fissi-, fissilis -is -e, fissuratus -a -um, fissus -a um cleft, divided

fistulosus -a -um hollow, pipe-like, tubular

flabellatus -a -um fan-like, fan-shaped

flabellifer -era -erum fan-bearing (with flabellate leaves) *flabelliformis -is -e* pleated fanwise

flaccidus -a -um limp, weak, feeble, flaccid

flaccus -a -um flabby, drooping

flagellaris -is -e, flagellatus -a -um with long thin shoots, whip-like, stoloniferous

flammeus -a -um fiery red, flame-red

flammula an old generic name for lesser spearwort

flammulus -a -um flame-coloured

flavens being yellow

flav-, flavi-, flaveolus -a -um yellowish

flavescens pale-yellow, turning yellow

flavidus -a -um yellowish

flavus -a -um bright almost pure yellow

flexi-, flexilis -is -e pliant, flexible

flexuosus -a -um zig-zag, winding, much bent, tortuous

-flexus -a um -turned

flocciger -era -erum, floccosus -a -um with a woolly indumentum which falls away in tufts, floccose

flore-albo white-flowered

florentinus -a -um from Florence, Florentine

flore-pleno double-flowered

floribundus -a -um abounding in flowers, freely flowering

floridanus -a -um from Florida, U.S.A.

floridus -a -um flowery, free-flowering

florindae for Florinda N. Thompson

florulentus -a -um flowery

-florus -a -um -flowered

flos-cuculi cuckoo-flowered, flowering in the season of cuckoo song

flos-jovis Jove's flower

fluctuans inconstant, fluctuating

fluitans floating on water

fluminensis -is -e growing in running water

fluvialis -is -e, fluviatilis -is -e growing in rivers and streams

foecundus -a -um very fruitful

foemina feminine

Foeniculum Latin for fennel

foenisicii of mown hay

foetidus -a -um, foetens stinking, bad-smelling

foliaceus -a -um leaf-like

foliatus -a -um leafy

foliosus -a -um leafy

-folius -a -um -leaved

88

follicularis -is -e bearing follicles (seed capsules as in hellebores)

fontanus -a -um, fontinalis -is -e of fountains, springs or fast-running streams

formicarius -a -um relating to ants

-formis -is -e -shaped

formosanus -a -um from Taiwan (Formosa)

formosus -a -um handsome, beautiful, well-formed

fornicatus -a -um arched

forrestii for George Forrest

fortis -is -e strong

fortunatus -a -um favourite, rich

fortunei for Robert Fortune

foulaensis -is -e from Foula, Shetland

foveolatus -a -um with small depressions or pits all over the surface

Fragaria fragrance (the fruit)

fragi- strawberry-

fragifer -era -erum strawberry-bearing

fragilis -is -e brittle, fragile

fragrans sweet-scented

Frangula breaking (the brittle twigs)

fraternus -a -um closely related, brotherly

fraxineus -a -um ash-like, resembling ash

Fraxinus ancient Latin name used in Virgil

frene- strap-

fresnoensis -is -e from Fresno County, California

frigidus -a -um cold, of cold habitats

friscus -a -um Friesian

Fritillaria dice box (the shape of the flower)

frondosus -a -um leafy

fructifer -era -erum fruit-bearing, fruitful

fructu- fruit-

frumentaceus -a -um grain-producing

frutescens, fruticans, fruticosus -a -um shrubby

fruticulosus -a -um dwarf-shrubby

fucatus -a -um painted, dyed

fucifer -era -erum drone-bearing

fuciformis -is -e, fucoides resembling bladder wrack (seaweed)

fugax fleeting, rapidly withering

fulgens, fulgidus -a -um shining, glistening (often of red flowers)

fuliginosus -a -um dirty brown to blackish, sooty

fullonum of cloth fullers

fulvescens becoming tawny

fulvi-, fulvus -a -um tawny, reddish-yellow

Fumaria smoke of the earth (medieval name referring to the smell of some, and hazy colour effect)

fumidus -a -um smoke-coloured, dull grey coloured

fumosus -a -um smoky

funebris -is -e funereal, mournful, doleful, of graveyards

fungosus -a -um spongy, fungus-like

funiculatus -a -um like a thin cord

furcans, furcatus -a -um forked, furcate

furfuraceus -a -um scurfy, mealy, scaly

furiens exciting to madness

fuscatus -a -um somewhat dusky-brownish

fusci-, fusco-, fuscus -a -um bright brown, swarthy, dark-coloured

fusiformis -is -e spindle-shaped

futilis -is -e useless

galactinus -a -um milky

galanthus -a -um milk-white flowered

Galanthus milk flower

galbanifluus -a -um with a yellowish exudate

galbinus -a -um greenish-yellow

gale from an old English vernacular name for bog myrtle or sweet gale

galeatus -a -um, galericulatus -a -um helmet-shaped, like a skull-cap

Galega milk-promoting

Galeobdolon a name used in Pliny (weasel-smelling)

Galeopsis an ancient Greek name (weasel-resembling)

galioides resembling *Galium*, bedstraw-like

Galium milk (the flowers of *G. verum*, ladies bedstraw, were used to curdle milk in cheesemaking)

gallicus -a -um from France, French

gamo- fused-, united-, married-

gandavensis -is -e from Ghent, Belgium

gangeticus -a -um from the Ganges region

garganicus -a -um from Mount Gargano, southern Italy

gaster-, gastro- belly-, bellied-

Gasteria belly (the swollen base of the corrolla)

Gastridium little paunch (the bulging of the glumes)

Gaura superb (the flowers)

geito-, geitono- neighbour-

gelidus -a -um of icy regions, growing in icy places

gemellus -a -um twinned, paired, in pairs

geminatus -a -um, gemini- twinned, united in pairs

gemmatus -a -um jewelled

gemmiferus -a -um, gemmiparus -a -um bearing buds or propagules

genavensis -is -e, genevensis -is -e from Geneva, Switzerland

generalis -is -e normal, usual

geniculatus -a -um with a knee-like bend

Genista a name in Virgil (*planta genista* from which the Plantagenets took their name)

Gentiana a name in Pliny for a King of Illyria

gentilis -is -e noble, of the same race, foreign

genuinus -a -um true, natural

geo- on or under the earth-

geocarpus -a -um with fruits which ripen underground

geoides *Geum*-like

georgei for George Forrest

georgianus -a -um from Georgia, U.S.A.

georgicus -a -um from Georgia, Caucasus, U.S.S.R.

Geranium crane (Dioscorides' name refers to the shape of the fruit resembling the head of a crane).

germanicus -a -um from Germany, German

-geton -neighbour

Geum a classical name in Pliny

gibb-, gibbosus -a -um swollen or enlarged on one side

gibberosus -a -um humped, hunchbacked

giganteus -a -um unusually large or tall

gigas giant

gileadensis -is -e of Gilead, Jordan

giluus -a -um dull yellow

gilvo-, gilvus -a -um dull pale yellow

gingidius -a -um from an old generic name used by Dioscorides

Ginkgo derived from a Japanese name

githago an old generic name (green with red-purple streaks)

glabellus -a -um somewhat smooth

glaber -ra -rum, glabro- smooth, without hairs, glabrous

glaberrimus -a -um very smooth

glabratus -a -um, glabrescens becoming smooth or glabrous

glabriusculus -a -um rather glabrous

glacialis -is -e of the ice, of frozen habitats

gladiatus -a -um sword-like

Gladiolus small sword (the leaves)

glandulifer -era -erum gland-bearing

glandulosus -a -um glandular

glaucescens, glaucus -a -um with a fine whitish bloom, bluish-green, sea-green, glaucous

glaucifolius -a -um with grey-green leaves

glauciifolius -a -um *Glaucium*-leaved

Glaucium grey-green (a name in Pliny)

Glaux a name used by Dioscorides

Gliricidia mouse killer (the poisonous seed and bark)

glischrus -a -um sticky, gluey

globatus -a -um arranged or collected into a ball

globosus -a -um, globularis -is -e spherical, with small spherical parts (e.g. flowers)

globulifer -era -erum carrying small balls (the sporocarps of pillwort)

globulosus -a -um small round-headed

glochidiatus -a -um with short barbed detachable bristles

glomeratus -a -um collected into heads, aggregated

glomerulatus -a -um with small clusters or heads

glosso-, glossus -a -um, -glottis tongue-shaped

glumaceus -a -um with chaffy bracts, conspicuously glumed

glutinosus -a -um sticky, viscous

Glyceria sweet

Glycine sweet (the roots of some species)

glyco-, glycy- sweet tasting or smelling

Glycyrrhiza sweet root (the source of liquorice)

glypto- cut into, carved

Gnaphalium soft down (the felted leaves)

gompho- nail-, bolt- or club-shaped

gongylodes roundish, knob-like, turnip-shaped, swollen

gonio-, -gonus -a -um angled-, prominently angled-

gorgoneus -a -um gorgon-like

gossypi- cotton-plant-like

gothicus -a -um from Gothland, Sweden

gracilior more graceful

gracilis -is -e slender, graceful

gracillimus -a -um most slender, most graceful

graecus -a -um Grecian, Greek

gramineus -a -um grass-like, grassy

graminis -is -e of grasses, grass-like

grammatus -a -um marked with raised lines or stripes

granadensis -is -e, granatensis -is -e either from Granada, Spain or from Colombia, South America (formerly New Granada)

grandi, grandis -is -e large, powerful, full-grown, showy

graniticus -a -um of granitic rocks

granulatus -a -um, granulosus -a -um tubercled, as though covered
   with granules
graph-, graphys- marked with lines-, as though written upon
grat-, gratus -a -um pleasing, graceful
gratianopolitanus -a -um from Grenoble, France
Gratiola agreeableness (medicinal effect)
graveolens strong-smelling, heavily scented
griseus -a -um bluish- or pearl-grey
Groenlandia for Johannes Groenland of Paris
groenlandicus -a -um from Greenland
grosse-, grossi-, grossus -a -um very large, thick, coarse
Grossularia from the French 'groseille' – gooseberry
grossularioides gooseberry-like
gruinus -a -um of cranes
grumosus -a -um broken into grains, granular, tubercled
guajava South American Spanish name
guianensis -is -e from Guiana, northern South America
guineensis -is -e from West Africa
gummifer -era -erum gum-producing
gummosus -a -um gummy
guttatus -a -um covered with small glandular dots, spotted
Gymnadenia naked gland (exposed viscidia of pollen)
gymno- naked-
gyno-, -gynus -a -um relating to the ovary, female, -carpelled
Gynura female tail (the stigma)
Gypsophila chalk lover (the natural habitat)
gyrans revolving, moving in circles
gyro- bent-, twisted-
gyrosus -a -um bent backwards and forwards

habr-, habro- soft-, delicate-, beautiful-
hadriaticus -a -um from the shores of the Adriatic
haema-, haemalus -a -um, haemorrhoidalis -is -e, haematodes
   blood-red, the colour of blood
Haemanthus blood flower (the fireball lilies)
haemanthus -a -um with blood-red flowers
halepensis -is -e, halepicus -a -um from Aleppo, northern Syria
halicacabus -a -um an ancient Greek name
halimi-, halimus -a -um orache-like, with silver-grey rounded
   leaves
Halimione daughter of the sea
halo-, halophilus -a -um salt-, salt-loving (the habitat)
hama- together with-

Hamamelis a Greek name for a tree with pear-shaped fruits

hamatus -a -um, hamosus -a -um hooked, hooked at the tip

Hammarbya for Linnaeus who had a house at Hammarby, Sweden

hamulatus -a -um having a small hook

hamulosus -a -um covered with little hooks

haplo- single-, simple-

harmalus -a -um adapting, responding, sensitive

harpe- sickle-

harpophyllus -a -um with sickle-shaped leaves

hastati-, hastatus -a -um formed like an arrow-head, spear-shaped (see Fig. 6(*a*))

hastifer bearing a spear

Hebe Greek goddess of youth, daughter of Jupiter

hebe- pubescent- (youthful)

hebecaulis -is -e slothful-stemmed (prostrate)

hebraicus -a -um Hebrew

hecisto- viper-like-

Hedera the Latin name for ivy

hederaceus -a -um, hederi- ivy-like, resembling *Hedera* (usually in leaves)

hedys sweet, of pleasant taste or smell

Hedysarum a name used by Dioscorides

Helenium for Helen of Troy (a name used by the Greeks for another plant)

heli-, helio- sun-

Helianthemum sun flower

Helianthus sun flower

Helichrysum golden sun

Helictotrichon (*um*) twisted hair (the awns)

helioscopius -a -um sun-observing, sun-watching

Heliotropium turn with the sun

helix ancient Greek name for twining plants

Helleborus the ancient name for the medicinal *H. orientalis*

hellenicus -a -um from Greece, Grecian, Greek

helo, helodes of bogs-, of marshes-

helodoxus -a -um glory of the marsh

helveticus -a -um from Switzerland, Swiss

helvolus -a -um pale yellowish-brown

helvus -a -um honey-coloured, dimly yellowish

Helxine a name used by Dioscorides, formerly for pellitory

Hemerocallis day beauty (the flowers are short-lived)

hemi- half-

hemidartus -a -um half-flayed, patchily covered with hair

hemionitideus -a -um barren, like a mule

Hemizonia half-embraced (the achenes)

Hepatica liver (signature of leaf or thallus shape as of use for liver complaints)

hepta- seven-

Heracleum Hercules' healer (a name used by Theophrastus)

herbaceus -a -um not woody, herbaceous, low-growing

hercoglossus -a -um coiled tongue

hercynicus -a -um from the Harz mountains, mid-Germany

Herminium buttress (the pillar-like tubers)

Hermodactylus Hermes' fingers

Herniaria rupture wort

herpeticus -a -um ringworm-like

Hesperis evening (Theophrastus for the evening flowering).

hespero- evening-, western-

heter-, hetero- diversely-, differing-

Heteranthera differing anthers (has one large and two small)

heterophyllus -a -um diversely leaved

hex-, hexa- six-

hexagonus -a -um six-angled

hexandrus -a -um six-stamened

hians gaping

hibernalis -is -e of winter (flowering or leafing)

hibernicus -a -um from Ireland, Irish

hibernus -a -um flowering or green in winter, Irish

Hibiscus an old Greek name for mallow

hiemalis -is -e of winter

Hieracium hawk (Dioscorides' name for the supposed use by hawks to give acute sight)

Hierochloe holy grass

hierochunticus -a -um from the classical name for Jericho (*Anastatica*, rose of Jericho)

Himantoglossum strap tongue

hippo- horse-

Hippocastanum horse chestnut

Hippocrepis horse shoe (the shape of the fruit)

Hippophae a name used by Theophrastus

Hippuris horse tail

hircinus -a -um of goats (the smell)

hirsutissimus -a -um very hairy

hirsutulus -a -um, hirtellus -a -um, hirtulus -a -um somewhat hairy *hirsutus -a -um* hairy, rough-haired

hirti-, hirtus -a -um hairy, shaggy hairy

hispalensis -is -e from Seville, southern Spain

hispanicus -a -um from Spain, Spanish

hispi-, hispidulus -a -um bristly, with stiff hairs

histrio of varied colouring, theatrical

histrionicus -a -um of actors, of the stage

histrix theatrical, showy

Holcus Greek for millet

hollandicus -a -um from northern New Guinea, from Holland

holo- completely-, entire-, entirely-

Holoschoenus a name used by Theophrastus

holosericus -a -um entirely silk-wrapped, completely covered in silk

Holosteum, holostea whole bone (an ancient Greek name for a chickweed-like plant)

homal-, homalo- smooth-, flat-

Homalocephala flat head (the tops of the flowers)

homo-, homoio- similar-, not varying-, agreeing with-

homolepis -is -e uniformly covered with scales

Hordeum Latin name for barley

horizontalis -is -e flat/on the ground, spreading horizontally

horminoides resembling clary

Horminum the Greek name for sage

hormo- chain-, necklace-

horridus -a -um very thorny, rough, sticking out

hortensis -is -e, hortorum, hortulanus -a -um of gardens, cultivated

hortulanorum of gardeners

hugonis for Fr Hugh Scallon

humifusus -a -um spreading over the ground, sprawling

humilis -is -e low-growing, smaller than most of its kind

Humulus from the Slavic-German 'chmeli'

hungaricus -a -um from Hungary, Hungarian

hupehensis -is -e form Hupeh, China

Hura from a South American name

Hyacinthus Homer's name for the flower which sprang from the blood of Hyakinthos, or from an earlier Thraco-pelasgian word for the blue colour of water

hyacinthus -a -um, hyacinthinus -a -um dark purplish-blue, resembling hyacinth

hyalinus -a -um nearly transparent, hyaline

hybernalis -is -e, hybernus -a -um of winter

hybridus -a -um bastard, mongrel, cross-bred, hybrid

Hydrangea water vessel (shape of the capsule)
Hydrilla water serpent
hydro- water-, of water-
Hydrocharis water beauty
Hydrochloa water grass
Hydrocotyle water cup
hydrolapathum a name in Pliny for a water dock
hydropiper water pepper
hyemalis -is -e of winter, winter (flowering)
hylaeus -a -um, hylo- of woods, of forests
hylophilus -a -um wood-loving
hymen-, hymeno- membrane-, membranous-
Hymenophyllum membranous leaf (filmy ferns)
Hyoscyamus hog bean (a derogatory name by Dioscorides)
hyper- above-, over-
hyperboreus -a -um of the far north
Hypericum above pictures (early use over shrines to repel evil spirits)
hypnoides resembling *Hypnum*, moss-like
hypo- under-, beneath-
Hypochoeris a name used by Theophrastus
hypochondriacus -a -um sombre, melancholy (colour)
hypochrysus -a -um golden underside, golden below
hypogaeus -a -um underground
hypophegeus -a -um from beneath beech trees
hypopithys, hypopitys growing under pine trees
hyrcanus -a -um from the Caspian Sea area
hyssopi- resembling hyssop
Hyssopus from a Semitic word 'ezob'
hystri, hystrix porcupine-like

ianthinus -a -um, ianthus -a -um bluish-purple, violet-coloured
-ias -much resembling
ibericus -a -um either from the Iberian peninsula (Spain and Portugal) or from the Georgian Caucasus
Iberis Dioscorides' name for an Iberian plant
-ibilis -is -e -able, -capable of
-icans -becoming, -resembling
-icola -of, -dwelling in
icos- twenty-
icosandrus -a -um twenty-stamened
ictericus -a -um jaundiced, yellowed
-icus -a -um -of, -from, -belonging to

idaeus -a -um from Mount Ida in Crete, or Mount Ida in N.W. Turkey

-ides, *(-oides)* -similar to, -resembling, -like

idoneus -a -um worthy, apt, suitable

ignescens, igneus -a -um fiery red

il-, im-, in- in-, into-, for-, contrary-, contrariwise-

Ilex the Latin name for the cork oak (*Quercus ilex*)

-ilis -is -e -able, -having, -like

Illecebrum a name in Pliny

illecebrosus -a -um alluring, enticing, charming

illinitus -a -um smeared, smudged

-illius -a -um -lesser (a diminutive ending)

illustratus -a -um pictured, painted, as if painted upon

illustris -is -e brilliant

illyricus -a -um from western Yugoslavia (Illyria)

ilvensis -is -e from Elba Island, from the river Elbe

imbecillis -is -e, imbecillus -a -um feeble, weak

imberbis -is -a without hair, unbearded

imbricatus -a -um overlapping (leaves, bracts, scales)

immaculatus -a -um without spots, unblemished, immaculate

immarginatus -a -um without a rim or border

immersus -a -um growing underwater

impari- unpaired-, unequal-

Impatiens impatient (touch-sensitive fruits)

impeditus -a -um tangled, hard to penetrate

imperator chief, leader

imperialis -is -e very noble

implexus -a -um tangled

imponens deceptive

impressus -a -um sunken, impressed (e.g. veins on a leaf)

impudicus -a -um lewd, shameless, impudent

in- not-

inaequidens with unequal teeth, not equally toothed

inapertus -a -um closed, not open

incanus -a -um quite grey, hoary-white

incarnatus -a -um made of flesh, flesh-coloured

incertus -a -um doubtful, uncertain

incisus -a -um sharply and deeply cut into

incomparabilis -is -e beyond compare, incomparable

incomptus -a -um unadorned

inconspicuus -a -um small

incubaceus -a -um lying close to the ground

incurvus -a -um curved, inflexed

indicus -a -um from India or, loosely, from the Orient
induratus -a -um hard (usually of an outer surface)
inebrians able to intoxicate, inebriating
inermis -is -e unarmed, without spines
-ineus -a -um -ish, -like
infaustus -a -um unfortunate
infectoreus -a -um dyed
infestus -a -um troublesome, hostile, dangerous
infirmus -a -um weak, feeble
inflatus -a -um swollen, inflated
inflexus -a -um bent in, curved inwards
infortunatus -a -um unfortunate (poisonous)
infra- below-
infractus -a -um curved inwards
infundibuliformis -is -e funnel-shaped, trumpet-shaped
ingens huge, enormous
innatus -a -um inborn, natural
innominatus -a -um unnamed, not named
innoxius -a -um without prickles, harmless
inodorus -a -um scentless, without smell
inophyllus -a -um fibrous-leaved
inopinatus -a -um, *inopinus -a -um* surprising, unexpected
inops deficient, poor
inornatus -a -um without ornament, unadorned
inquillinus -a -um introduced
inquinans stained, blemished
inscriptus -a -um as though written upon, inscribed
insectifer -era -erum bearing insects (mimetic fly orchid)
insertus -a -um inserted (the scattered inflorescences)
insignis -is -e remarkable, striking, decorative
insititius -a -um grafted
insubricus -a -um from the Lapontine Alps (Insubria) between
    Lake Maggiore and Lake Lucerne
insulanus -a -um, *insularis -is -e* growing on islands, insular
intactus -a -um untouched, unopened (the flowers)
integer -gra, *-grum*, *integerrimus -a -um*, *integri-* undivided, entire,
    intact
integrifolius -a -um with entire leaves
inter- between-
interjectus -a -um intermediate in form, interposed
intermedius -a -um between extremes, intermediate
interruptus -a um with scattered leaves or flowers
intertextus -a -um interwoven

intortus -a -um twisted

intra- within-

intricatus -a -um entangled

introrsus -a -um facing inwards, turned towards the axis

intumescens swollen

intybus from a name in Virgil for wild chicory or endive

Inula the Latin name, in Pliny

inundatus -a -um flooded, of marshes, of places which flood

-inus -a -um -ish, -like, -from

invenustus -a -um lacking charm, unattractive

inversus -a -um turned over, inverted

involucratus -a -um surrounded with bracts (the flowers), with an involucre, involucrate

involutus -a -um rolled inwards, involute, obscured

iodes resembling *Viola*, violet-like

ioensis -is -e from Iowa, U.S.A.

ion-, iono- violet

-ion -occurring

ionantherus -a -um with violet-coloured flowers, violet-flowered

ionanthus -a -um with violet-coloured flowers

ionicus -a -um from the Ionian Islands, Greece

iricus -a -um from Ireland, Irish

iridescens iridescent

iridi- *Iris*-like-

irio an ancient Latin name for a cruciferous plant

Iris the name of the mythical goddess of the rainbow

irrigatus -a -um of wet places, flooded

irriguus -a -um watered

irritans causing irritation

isabellinus -a -um yellowish, tawny

isandrus -a -um with equal stamens

Isatis the name used by Hippocrates for woad

Ischaemum blood stopper (a name in Pliny for styptic property)

-iscus -a -um -lesser (diminutive ending)

islandicus -a -um from Iceland

iso- equal-

Isoetes equalling one year (green throughout)

Isolepis equal scales (the glumes)

Isoloma equal lobes (of the perianth)

-issimus -a -um -est, -the best, -the most (superlative)

istriacus -a -um from Istria, Yugoslavia

-ium -lesser (diminutive ending)

italicus -a -um from Italy, Italian

100

iteophyllus -a -um willow-leaved
ites, -itis -closely resembling, -very much like
ixocarpus -a -um sticky-fruited

jacea from the Spanish name for a knapweed
jacobaeus -a -um either for St James (Jacobus) or from Iago
    Island, Cap Verde
Jacobinia from Jacobina, South America
jalapa from a South American Indian name
jambos a Malaysian name
japonicus -a -um from Japan, Japanese
Jasione healer (the Greek name for another plant)
Jasminum from the Persian 'yasmin'
javanicus -a -um from Java, Javanese
jejunus -a -um meagre, small
johannis from Port St John, South Africa
juanensis -is -e of Genoa, northern Italy
jubatus -a -um maned, crested with awns
jucundus -a -um pleasing
jugalis -is -e yoked, joined together
Juglans Jupiter's nut
juliae for Julia Mlokosewitsch who discovered *Primula juliae*
juliformis -is -e downy
junceus -a -um, juncei-, junci- rush-like, resembling *Juncus*
Juncus binder (classical Latin name refers to use for weaving
    and basketry)
Juniperus the Latin name

kaido a Japanese name
Kalanchoe a Chinese name
kali either from the Persian for a carpet, or a reference to the
    ashes of saltworts being alkaline ('alkali')
Kalmia for Peter Kalm, a highly reputed student of Linnaeus
kalo- beautiful-
Kalopanax beautiful *Panax*
kamtschaticus -a -um from the Kamchatka peninsula, eastern
    U.S.S.R.
Kentranthus (*Centranthus*) spur flower
kermesinus -a -um carmine-coloured, carmine
kewensis -is -e of Kew Gardens
khasianus -a -um from the Khasi Hills, Assam
kirro- citron-coloured
kisso- ivy-

101

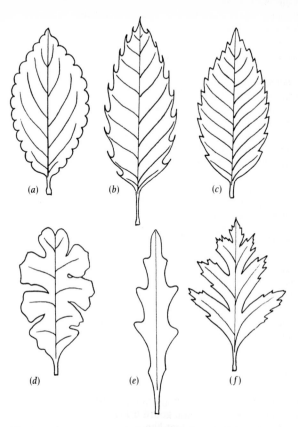

Fig. 4. Leaf-margin features which provide specific epithets:
(a) crenate (scalloped as in *Ardisia crenata* Sims); (b) dentate (toothed
as in *Castanea dentata* Borkh.). This term has been used for a range
of marginal tooth shapes; (c) serrate (saw-toothed as in *Zelkova
serrata* (Thunb.) Makino); (d) lobate (lobed as in *Quercus lobata* Née);
(e) sinuate (wavy as in *Matthiola sinuata* (L.) R.Br.). This refers to 'in
and out' waved margins and not 'up and down' or undulate waved
margins; (f) laciniate (cut into angular segments as in *Crataegus
laciniata* Ucria).

kobus a Japanese name
Kolomicta from a vernacular name from Amur, eastern U.S.S.R.
koreanus -a -um, koriensis -is -e from Korea, Korean
kousa a Japanese name for a *Cornus* species
kurroo from a Himalayan name

labiatus -a -um lip-shaped
labilis -is -e unstable, labile
labiosus -a -um conspicuously lipped
Laburnum the name in Pliny
-lacca -resin
lacciferus -a -um producing a milky juice
lacer -era -erum, laceratus -a -um torn into a fringe, looking torn
lachno- woolly-, downy-
laciniatus -a -um, laciniosus -a -um jagged, unevenly cut, slashed
  (see fig. 4(*f*))
lacistophyllus -a -um having torn leaves
lacrimans weeping, causing tears
lacryma-jobi Job's tears
lactescens having lac or milk
lacteus -a -um, lact-, lacti- milk-coloured
lactifer -era -erum producing a milky juice
Lactuca the Latin name (has milky juice)
lacunosus -a -um with gaps, pits, furrows or deep holes
lacuster -tris -tre, lacustris -is -e of lakes or ponds
ladanifer -era -erum bearing ladanum (the resin called myrrh)
laetevirens bright-green
laeti-, laetis -is -e, laetus -a -um pleasing, bright
laevi-, laevigatus -a -um, laevis -is -e polished, smooth, not rough
lag-, lago- hare's-
lagaro-, lagaros- lanky-, long-, narrow-, thin-
Lagenaria flask (the bottle gourd fruit)
lagopinus -a -um hare's-foot-like
lagopus -a -um hare's foot
Lagurus hare's tail
Lamarckia for Jean Baptiste Antoine Pierre Monnet de Lamarck,
  evolutionist
lamellatus -a -um layered, lamellated
lamii- deadnettle-like, resembling *Lamium*
Lamiopsis looking like *Lamium*
Lamium gullet (the name in Pliny refers to the shape of the
  corolla tube)
lampro- shining-, glossy-

lanatus -a -um woolly

lanceolatus -a -um, lanci- narrowed and tapering at both ends

lanceus -a -um spear-shaped

landra from the Italian name for a radish

laniger -era -erum, lanosus -a -um, lanuginosus -a -um softly hairy, woolly or cottony

Lantana an old name for *Viburnum*

lapathi- sorrel-like, dock-like-

lappa-, lappaceus -a -um bearing buds, bud-like

lapponicus -a -um, lapponus -a -um from Lapland, of the Lapps

lappulus -a -um with small burs (the nutlets)

Lapsana Dioscorides' name for a salad plant

laricinus -a -um larch-like, resembling *Larix*

laricio the Italian name for several pines

Larix Dioscorides' name for a larch

lasi-, lasio- woolly-, shaggy-

lasiolaenus -a -um shaggy-cloaked

latebrosus -a -um of dark or shady places

lateralis -is -e, lateri- on the side, laterally

latericius -a -um, lateritius -a -um brick-red

Lathraea hidden (fairly inconspicuous parasites)

lathyris the Greek for a kind of spurge

Lathyrus the ancient name for the chickling pea

lati-, latus -a -um broad, wide

latifrons with broad fronds

latipes broad-stalked, thick-stemmed

latobrigorum of the Rhinelands

laureolus -a -um of garlands

lauri- laurel-

lauricatus -a -um wreathed, resembling laurel or bay

laurocerasus -a -um cherry-laurel

Laurus the Latin name for laurel or bay

lautus -a um washed

lavandulae- lavender-

laxi, laxus -a -um open, loose, not crowded, distant, lax

lazicus -a -um from N.E. Turkey (Lazistan)

lecano- basin-

Lecythis oil jar (the shape of the fruit)

Ledum an old Greek name for a rockrose

legionensis -is -e from León, Spain

leio- smooth-

Lemna Theophrastus' name for a water plant

lendiger -era -erum nit-bearing (the appearance of the spikelets)

Lens the classical name for the lentil
lenticularis -is -e lens-shaped, bi-convex
lenticulatus -a -um with conspicuous lenticels on the bark
lentiformis -is -e lens-shaped, bi-convex
lentiginosus -a -um freckled, mottled
lentus -a -um tough, pliable
leodensis -is -e from Liège, Belgium
leonensis -is -e from Sierra Leone, West Africa
leonis -is -e toothed or coloured like a lion
Leonotis lion's ear
leonto- lion's-
Leontodon lion's tooth
Leonurus lion's tail
Lepidium little scale (Dioscorides' name for a cress refers to the
   fruit)
lepido- scaly-, flaky-
lepidotus -a -um scurfy, scaly
lepidus -a -um neat, elegant, graceful
-lepis -scaly, -scaled
leporinus -a -um hare-like
lept- hare-like-, slender-
lepta-, lepto- slender-, weak-, small-, thin-, delicate-
leptochilus -a -um with a slender lip
leptophyllus -a -um slender-leaved
Lepturus hare's tail
leuc-, leuco- white-
Leucanthemum white flower (Dioscorides' name)
leuce a name for the white poplar
Leucojum white violet (Hippocrates' name for a snowflake)
levigatus -a -um smooth, polished
levis -is -e smooth, not rough
libanensis -is -e, libanoticus -a -um from Mount Lebanon
libani from the Lebanon, Lebanese
libanotis -is -e from Mount Lebanon, or of incense
libericus -a -um from Liberia, West Africa
libero- bark-
liburnicus -a -um from Croatia (Liburnia) on the Adriatic
libycus -a -um from Libya, Libyan
lignescens turning woody
lignosus -a -um woody
ligtu from a Chilean name
Ligularia strap (the shape of the ray florets)
ligulatus -a -um strap-shaped

Ligusticum the name of a plant from Liguria, N.E. Italy
Ligustrum a name used in Virgil
lilacinus -a -um lilac-coloured, resembling lilac
lili- lily-
liliaceus -a -um lily-like, resembling *Lilium*
liliago the silvery
Lilium the name in Virgil
lilliputianus -a -um of very small growth, Lilliputian
limaeus -a -um of stagnant waters
limbatus -a -um bordered, with a margin
limbo- border-, margin-
limensis -is -e from Lima, Peru
Limnanthemum pond flower
Limnanthes marsh flower
limno- marsh-, pool-, pond-
limnophilus -a -um marsh-loving
Limonium Dioscorides' name for a meadow plant
Limosella muddy
limosus -a -um muddy, slimy, of muddy places
lin-, lini- flax-
Linaria flax-like (the leaves)
linarii- flax-like-
linearis -is -e narrow and parallel-sided (usually leaves)
lineatus -a -um marked with lines (usually parallel and coloured)
lingulatus -a -um, linguus -a -um tongue-shaped (*Linguus* was a
    name used in Pliny)
linicolus -a -um of flax fields
Linnaea, linnaeanus -a -um, linnaei for Carl Linnaeus
linosyris flax-flowered, an old generic name
Linum the ancient Latin name for flax
liolaenus -a -um smooth-cloaked, glabrous
Liparis greasy (the leaf texture)
lirio- lily-white-
Liriodendron lily tree (the showy flowers of the tulip tree)
lisso- smooth-
literatus -a -um with the appearance of being written upon
litho- stone-
lithophilus -a -um stone-loving, living in stony places
Lithops stone-like (appearance of stone cacti)
Lithospermum stone seed (the texture of the nutlets)
lithuanicus -a -um from Lithuania, Lithuanian
litoralis -is -e, littoralis -is -e, littorius -a -um of the sea-shore,
    growing by the shore

Littorella shore (the habitat)

lividus -a -um lead-coloured, grey, bluish-brown

lizei for the Lizé Frères of Nantes

lobatus -a -um, lobi-, lobus -a -um with lobes, lobed (see Fig. 4(*d*))

lobius -a -um, -lobion -pod, -podded

Lobularia small pod

lochabrensis -is -e from Lochaber, Scotland

lochmius -a -um coppice-dweller, of thickets

locustus -a -um spikeleted

loganobaccus -a -um after J. H. Logan who developed the logan-
  berry

loliaceus -a -um resembling *Lolium*

Lolium a name in Virgil for a weed grass

-loma -fringe, -border

Lomaria border (the marginal sori)

lonchitis -is -e, loncho- spear-shaped, lance-shaped (a name used
  by Dioscorides for a fern)

longe-, longi-, longus -a -um long

longipes long-stalked

lophanthus -a -um with crested flowers

lopho- crest-, crested-

Lophophora crest bearer (the tufts of glochidiate hairs)

Loranthus strap flower (from the shape of the 'petals')

loratus -a -um, lori-, loro- strap-shaped

loti-, lotoides Lotus-like

Lotus the ancient name was applied to several plants

louisianus -a -um from Louisiana, U.S.A.

loxo- oblique-

lucens, lucidus -a -um glittering, shining, clear

luciliae for Lucile Boissier

lucorus -a -um of woods or woodland

ludovicianus -a -um from Louisiana, U.S.A.

Luffa from the Arabic name

Lunaria moon (the shape and colour of the septum, or repla, of
  the fruit of honesty)

lunatus -a -um, lunulatus -a -um half-moon shaped

Lupinus the ancient Latin name of the white lupin

lupuli-, lupulinus -a -um hop-like, resembling *Humulus*

lupulus Wolf (the ancient Latin name for hop was a reference to
  its habit of straggling over other plants – willow wolf)

luridus -a -um sallow, dingy yellow or brown

lusitanicus -a -um from Portugal (Lusitania), Portuguese

lutarius -a -um of muddy places, living on mud

luteo-, *luteus -a -um* yellow
luteolus -a -um yellowish
lutescens turning yellow
lutetianus -a -um from Paris (Lutetia), Parisian
luxurians luxuriant
Luzula ancient name of obscure meaning
Lychnis lamp (the hairy leaves were used as wicks)
lychnitis lamp (a name used in Pliny)
Lycium the name of a thorn tree from Lycia
lycius -a -um from Lycia, S.W. Turkey
lyco- wolf-
Lycopersicum wolf-peach (tomato)
Lycopus wolf's foot
Lycopodium wolf's foot (clubmoss)
lycotonum wolf-murder (for a very poisonous Aconite)
lydius -a -um from Lydia, S.W. Turkey
Lygodium willow-like (climbing fern stems)
lyratus -a -um lyre-shaped (rounded above with small lobes
    below – usually of leaves)
lysi-, *lysio-* loose-
Lysimachia ending strife (named after a Thracian king)
Lythrum blood (Dioscorides' name refers to the flower colour of
    some species)

macedonicus, -a -um from Macedonia, Macedonian
macellus -a -um rather meagre, poorish
macer -era, -erum meagre
macilentus -a -um thin, lean
macr-, *macro-* big-, large-, long-
macrodus -a -um large-toothed
macrurus -a -um long-tailed
maculatus -a -um, *maculi-*, *maculifer -era -erum* spotted, blotched,
    bearing spots
madagascariensis -is -e from Madagascar
maderaspatanus -a -um, *maderaspatensis -is -e* from the Madras
    region of India
maderensis -is -e from Madera
Madia from a Chilean name
madrensis -is -e from the Sierra Madre, northern Mexico
madritensis -is -e from Madrid, Spain
magellanicus -a -um from the Straits of Magellan, South America
magellensis -is -e from Monte Majella, Italy
magni-, *magno-*, *magnus -a -um* large, great

magnificus -a -um great, magnificent

mahaleb an Arabic name

mai-, maj- May-

Maianthemum May flower (a may flowering lily)

majalis -is -e of May, produced in May (e.g. flower)

majesticus -a -um majestic

major, majus -a -um larger, greater, bigger

malabaricus -a -um from the Malabar coast, S. India

malaco-, malako-, malacoides mucilaginous, soft, tender, weak

Malaxis softener (soft leaves)

maliformis -is -e apple-shaped

Malope a name in Pliny for mallow

Malus the Latin name for an apple tree

Malva the name in Pliny

malvinus -a -um mauve, mallow-like

mammaeformis -is -e, mammiformis -is -e nipple-like

mammilatus -a -um, mammilaris -is -e, mammosus -a -um having
nipples, with nipple-like structures

mandibularis -is -e jaw-like, having jaws

Mandragora Greek name derived from a Syrian one

mandschuricus -a -um from Manchuria, Manchurian

manicatus -a -um with a felty covering which can be stripped off,
with long sleeves

manipuliflorus -a -um with few-flowered clusters

manriqueorus -a -um for Manrique de Lara

mantegazzianus -a -um for Paulo Mantegazzi, Italian traveller

Manzanilla from the Spanish for a small apple

marcescens not putrifying, persisting, retaining dead leaves

marckii for J. B. A. P. Monnet de la Marck; Lamarck

margaritaceus -a -um, margaritus -a -um pearly

marginalis -is -e, marginatus -a -um having a distinct margin (the
leaves)

marianus -a -um of St Mary, from Maryland, U.S.A., or from the
Sierra Morena

marinus -a -um growing by or in the sea, marine

mariscus the name in Pliny for a rush

maritimus -a -um growing by the sea, maritime

marmelos a Portuguese name

marmoratus -a -um, marmoreus marbled

martagon resembling a kind of Turkish turban

martinicensis -is -e from Martinique

mas, maris, masculus -a -um male, bold

massiliensis -is -e from Marseilles, France

Matricaria of the womb (former medicinal use)

matritensis -is -e from Madrid, Spain

matronalis -is -e of the matrons (their Roman festival was held on March 1st)

mauritanicus -a -um from Morocco or North Africa generally

mauritianus -a -um from the Island of Mauritius, Indian Ocean

maurorum of the Moors, Moorish, of Mauritania

maxillaris -is -e of jaws

maximus -a -um largest, greatest

mays from the Mexican name for Indian corn

Meconopsis poppy-like

Medicago from a Persian name for a grass

medicus -a -um from Media, curative, medicinal

medio-, medius -a -um middle-sized, in between

mediopictus -a -um with a (coloured) stripe down the centre-line

mediterraneus -a -um from the Mediterranean region, from well inland

medullaris -is -e, medullosus -a -um, medullus -a -um pithy, soft-wooded with a large pith

mega-, megalo- big-, great-, large-

megalurus -a -um large-tailed

megapotamicus -a -um of the big river, from the Rio Grande or the Amazon river

megaseifolius -a -um Bergenia (Megasea)-leaved

mela-, melan-, melano- black-

Melaleuca black-white (bark colours on trunk and branch)

Melampyrum black wheat (a name used by Theophrastus)

melancholicus -a -um sad-looking, drooping

Melandrium the name used in Pliny

melanops black-eyed

Melastoma black mouth (the fruits stain)

meleagris -is -e a Greek name for Meleagris of Calydon, or chequered like a guinea fowl

Melica honey grass

Melilotus Theophrastus' name refers to melilot's attractiveness to bees

Melissa bee (their attraction to the flowers)

melissophyllus -a -um Melissa-leaved, balm-leaved

melitensis -is -e from Malta, Maltese

Melittis bee (bastard balm attracts bees)

mellifer -era -erum honey-bearing

mellitus -a -um honey-sweet, darling

melo- melon-

membranaceus -a -um thin in texture, skin-like, membranous

Mentha the name in Pliny for mint

Menyanthese moon flower (Theophrastus' name for *Nymphoides*)

Mercurialis named by Cato for Mercury, messenger of the gods

meridianus -a -um, *meridionalis -is -e* of noon, flowering at mid-day

-merus -a -um -partite, -divided into, -merous

mes-, *meso-* middle-

mesoponticus -a -um from the middle sea (lakes of central Africa)

mesopotamicus -a -um from between the rivers

Mespilus Theophrastus' name for a medlar

messaniensis -is -e from Messina, Italy

messeniensis -is -e from Messenia, Morea, Greece

met-, *meta-* amongst-, next to-, after-, behind-, later-

metallicus -a -um metallic in appearance

methysticus -a -um intoxicating

mexicanus -a -um from Mexico, Mexican

micaceus -a -um from mica soils

micans shining, glittering

micr-, *micro-* small-

microdon small-toothed

Microglochin small point

mikanioides resembling *Mikania* (climbing hempweed)

miliaceus -a -um millet-like

militaris -is -e soldier-like, resembling part of a uniform

mille- a thousand (loosely)

millefolius -a -um thousand-leaved (much divided leaves of milfoil)

mimetes mimicking

Mimosa mimic (the sensitivity of the leaves)

Mimulus ape-flower

miniatus -a -um red-lead-coloured, cinnabar-red

minimus -a -um least, smallest

minor -or -us small

minutissimus -a -um extremely small

minutus -a -um small

mirabilis -is -e astonishing, extraordinary, wonderful

mirandus -a -um extraordinary

missouriensis -is -e from Missouri, U.S.A.

mitis -is -e gentle, mild, bland, not acid (without spines)

mitratus -a -um mitred, turbaned

mitriformis -is -e mitre-shaped

mixtus -a -um mixed

111

modestus -a -um modest

moesiacus -a -um from the Balkans (Moesia)

moldavicus -a -um from the Danube area (Moldavia)

molinae for J. I. Molina

Molium magic garlic (after *Allium moly*)

molle a Peruvian name

molliaris -is -e supple, graceful, pleasant

mollis -is -e soft

Mollugo soft (a name in Pliny)

moluccanus -a -um from Indonesia (Moluccas)

moly Greek for a magic herb

mona-, mono- one-, single-, alone-

monensis -is -e from Anglesey or from the Isle of Man

moniliformis -is -e necklace-like, like a string of beads

Monotropa one turn (the band at the top of the stem)

monspeliensis -is -e, monospessulanus -a -um from Montpellier
    S. France

monstrosus -a -um abnormal, monstrous

montanus -a -um, monticolus -a -um of mountains, mountain
    dweller

morio madness

-morphus -a -um -shaped, -formed

morsus-ranae frog's mouth (frogbit)

mortuiflumis -is -e of dead water, growing in stagnant water

Morus the ancient Latin name for the mulberry

mosaicus -a -um coloured like a mosaic

moschatus -a -um musk-like, musky

moupinensis -is -e from Mupin, W. China

mucosus -a -um slimy

mucronatus -a -um with a hard sharp-pointed tip, mucronate (see
    Fig. 7(*b*))

multi-, multus -a -um many

mume from the Japanese 'ume'

mundus -a -um clean, neat, elegant, handsome

munitus -a -um armed

muralis -is -e of walls, growing on walls

muricatus -a -um rough with short superficial tubercles

murinus -a -um of mice, mouse-grey

murorum of walls

Musa from the Arabic name

musaicus -a -um mottled like mosaic

Muscari musk-like (from the Turkish – fragrance)

muscifer -era -erum fly-bearing

muscipulus -a -um fly-catching
muscivorus -a -um fly-eating
muscoides fly-like
muscosus -a -um moss-like
musi- banana-
mutabilis -is -e changeable
mutatus -a -um changed, altered
muticus -a -um without a point, not pointed, blunt
myagroides resembling *Myagrum*
Myagrum mouse-trap (Dioscorides' name)
myiagrus -a -um fly-catching (sticky)
Myosotis mouse ear (Dioscorides' name)
Myosoton mouse ear (Dioscorides' name)
Myosurus mouse tail
Myrica the ancient name for a tamarisk
Myriophyllum numerous leaves (Dioscorides' name)
myrmecophilus -a -um ant-loving (plants with special ant
    associations)
Myrrhis from the ancient name for true myrrh
myrsinites myrtle-like
mystacinus -a -um moustached
mysurensis -is -e from Mysore, India
myuros mouse-tailed
Myurus mouse tail

Naias, Najas a water nymph
namaquensis -is -e from Namaqualand, western South Africa
nana, nanae, nani, nano-, nanoe-, nanus -a -um dwarf
nanellus -a -um very dwarf
nannophyllus -a -um small-leaved
napaulensis -is -e from Nepal, Nepalese
napellus -a -um swollen, turnip-rooted, like a small turnip
napi- turnip-
Napus the name in Pliny for a turnip
narbonensis -is -e from Narbonne, southern France
Narcissus the name of a youth in Greek mythology, torpid (the
    narcotic effect)
Nardus spikenard (-like)
narinosus -a -um broad-nosed
Narthecium little rod (the stem)
Nasturtium nose twist (the mustard-oil smell)
natans floating under water
nauseosus -a -um nauseating

navicularis -is -e boat-shaped

neapolitanus -a -um from Naples

nebrodensis -is -e from Mt Nebrodi, Sicily

nebulosus -a -um cloud-like

neglectus -a -um (previously) disregarded, overlooked, neglected

negundo from a Sanskrit name

nelumbo from the Sinhalese name

nema-, -nema, nemato- thread-like-

Nemophila glade-loving (woodland habitat)

nemoralis -is -e, nemorosus -a -um, nemorus -a -um of shady
   groves, of woodland, sylvan

neo- new-

neomontanus -a -um from Neuberg, Germany

Neottia nest of fledglings

nepalensis -is -e from Nepal, central Himalayas

Nepenthes euphoria (its reputed drug property)

nepeta from Nepi, Italy

nephr-, nephro- kidney-

Nephrolepis kidney scale (the shape of the indusia of the sori)

nericus -a -um the Närke, Sweden

nerii—, Nerium from the ancient Greek name for oleander

nerterioides resembling *Nertera* (bead plants)

nervatus -a -um, nervis -is -e nerved or veined

nervosus -a -um with conspicuous nerves or veins

nesophilus -a -um island-loving

nessensis -is -e from Loch Ness, Scotland

-neurus -a -um -nerved, -veined

nicaeensis -is -e from Nice, S.E. France, or from Nicaea, Bithynia,
   N.W. Turkey

Nicandra for Nikander of Calophon, writer on plants of 100 B.C.

Nicotiana for Jean Nicot who introduced tobacco to France in
   the late sixteenth century

nictitans blinking, moving

nidus -a -um nest-like

nidus-avis bird's-nest

Nigella blackish (the seed coats)

niger, nigra, nigrum black

nigrescens, nigri-, nigro-, nigricans blackish, darkening, turning
   black

nikoensis -is -e from Nike, Japan

niliacus -a -um from the River Nile

niloticus -a -um from the Nile valley

nipho- snow-

nipponicus -a -um from Japan, Japanese

nitens, nitidi-, nitidus -a -um shining, glossy, with a polished surface, neat

nivalis -is -e snow-white, growing near snow

niveus -a -um, nivosus -a -um snow-white, purest white

nobilis -is -e noble, grand, famous

nocti- night-

noctiflorus -a -um, nocturnus -a -um night-flowering

nodiflorus -a -um flowering at the nodes

nodosus -a -um many jointed, conspicuously jointed, knotty

nodulosus -a -um noduled, with swellings (e.g. on roots)

Nolana small bell

noli-tangere touch (me) not (the ripe fruit ruptures on touch)

noma-, nomo- meadow-

nominius -a -um customary

non- not-, un-

nonscriptus -a -um not written upon, unmarked

nootkatensis -is -e from Nootka Sound, British Columbia

norvegicus -a -um from Norway, Norwegian

notatus -a -um spotted

notho-, nothos-, nothus -a -um false-, bastard, spurious

noti-, notio- southern-

noto- the back-, surface-

noveboracensis -is -e from New York, U.S.A.

novae-angliae from New England, U.S.A.

novae-zelandiae from New Zealand

novi-belgae from New York, U.S.A.

novi-caesareae from New Jersey, U.S.A.

nubicolus -a -um, nubigenus -a -um, nubilus -a -um of cloudy places

nubicus -a -um from the Sudan, N.E. Africa

nucifer -era -erum nut-bearing

nudatus -a -um, nudi-, nudus -a -um naked, bare

nudicaulis -is -e naked-stemmed, leafless

numidicus -a -um from Algeria (Numidia)

nummularis -is -e circular, coin-like (leaves)

nummularius -a -um moneywort-like, resembling *Nummularia*

Nuphar the Persian name for a water lily

nutans nodding, dropping (the flowers)

nutkanus -a -um as for *nootkatensis*

nycticalus -a -um, nyctagineus -a -um night-flowering

Nymphaea Nymphe (Theophrastus' name after one of the three water nymphs)

Nyssa Nyssa (another water nymph)

115

ob-, oc-, of-, op- contrary-, inverted-, inversely-, against-

obconicus -a -um like an inverted cone

obesus -a -um succulent, fat

obfuscatus -a -um clouded over, confused

oblatus -a -um oval, somewhat rounded at the ends

obliquus -a -um slanting, unequal-sided

oblongatus -a -um, oblongus -a -um elliptic with blunt ends

obovatus -a -um egg-shaped in outline and with the narrow end lowermost

obscurus -a -um dark, darkened, obscure

obsoletus -a -um rudimentary

obtectus -a -um covered over

obtusatus -a -um, obtusi-, obtusus -a -um blunt, rounded, obtuse

obvallaris -is -e, obvallatus -a -um walled around, enclosed

occidentalis -is -e western

occultus -a -um hidden

oceanicus -a -um growing near the sea

ocellatus -a -um eye-like, with a colour-spot bordered with another colour

ochraceus -a -um ochre-coloured

ochroleucus -a -um buff-coloured, yellowish-white

ocimoides resembling *Ocimum*, like sweet basil

Ocimum the Greek name for an aromatic plant

octa-, octo- eight-

octandrus -a -um eight-stamened

oculatus -a -um with an eye

ocymoides resembling *Ocimum*

-odes -like, -resembling

odessanus -a -um from Odessa, U.S.S.R.

Odontites for teeth (the name in Pliny refers to its use for treating toothache)

odonto- tooth-

odoratus -a -um, odorifer -era -erum, odorus -a -um fragrant, scented

oedo- becoming swollen-

Oenanthe wine-scented

Oenothera ass catcher (the Greek name for another plant)

officinalis -is -e, officinarus -a -um of shops, sold in shops, officinal medicines

-oides, -oideus -a -um -resembling, -like

olbia, olbios rich or from Hyères (Olbia), France

oleander from the Italian (for the olive-like leaves)

116

olei- olive-

oleifer -era -erum oil-bearing

-olentus -a -um -abundance, -fullness of

oleraceus -a -um of cultivation, suitable for food, vegetable, aromatic

olidus -a -um stinking, smelling

oligo- small-, feeble-, few-

olitorius -a -um culinary, of market gardens, salad vegetable

olusatrus -a -um Pliny's name for a black-seeded pot-herb

olympicus -a -um from Mt Olympus

omeiensis -is -e from Mt Omei, China (Szechwan)

omphalo- navel-

Omphalodes navelled (the fruits)

onco- tumour-, hook-

onites a name used by Dioscorides (of an ass or donkey)

Onobrychis a name in Pliny for a legume eaten greedily by asses

Onoclea closed cups (the sori are concealed by the rolled fronds)

Ononis the classical name used by Dioscorides

Onopordon ass fart

Onosma ass smell (said to attract asses)

oo- egg-shaped-

opacus -a -um dull, not glossy, shady

ophio- snake-

ophioglossifolius -a -um snake's-tongue-leaved

Ophioglossum snake tongue

Ophrys eyebrow (the name in Pliny)

opistho- behind-, back-

oporinus -a -um autumnal, of late summer

oppositi- opposed-, opposite-

-ops, -opsis -is -e -like, -looking like, -appearance of

opuli- guelder rose-like-

opulus an old generic name for the guelder rose

orarius -a -um of the shoreline

orbicularis -is -e, orbiculatus -a -um disc-shaped, circular in outline, orbicular

orcadensis -is -e from the Orkney Isles

Orchis testicle (the shape of the root-tubers)

oreganus -a -um, oregonus -a -um from Oregon, U.S.A.

oreo-, ores-, ori- mountain-

oreodoxa mountain glory

oreophilus -a -um mountain-loving

oresbius -a -um living on mountains

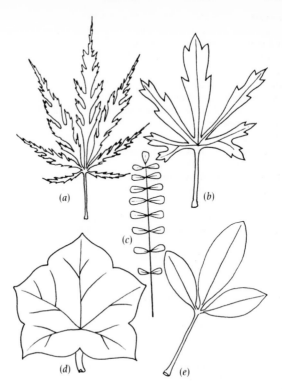

Fig. 5. Some leaf shapes which provide specific epithets:
(a) palmate (e.g. *Acer palmatum* Thunb. 'Dissectum'. As this maple's
leaves mature, the secondary division of the leaf lobes passes
through incised (*incisum*) to torn (*laciniatum*) to dissected), lobed
from one central point; (b) pedate (e.g. *Callirhoe pedata* Gray) is
distinguished from palmate by having the lower lobes themselves
divided; (c) pinnate (e.g. *Ornithopus pinnatus* Druce). When the lobes
are more or less strictly paired it is called paripinnate, when there
is an odd terminal leaflet it is called imparipinnate, and when the
lobing does not extend to the central leaf-stalk it is called pinnatifid;
(d) peltate (e.g. *Pelargonium peltatum* (L.) Ait) has the leaf-stalk
attached on the lower surface, not at the edge; (e) ternate (e.g.
Choisya ternata H.B.K.). In other ternate leaves the three divisions
may be further divided, ternately, palmately or pinnately.

118

organensis -is -e from Organ Mt, New Mexico, U.S.A. or Brazil

orgyalis, -is -e about 6 feet in length (the distance from finger-tip to finger-tip with arms stretched)

orientalis -is -e eastern, oriental

Origanum Theophrastus' name for an aromatic herb

-orius -a -um -capable of, -able, -functioning

ormo- necklace-, necklace-like-

ornatus -a -um, ornus -a -um from the ancient Latin for ash, showy, adorned

ornitho- bird-, bird-like-

Ornithogalum bird milk (yields a bird-lime)

ornithopodus -a -um like a bird's foot

Ornithopus bird foot (the fruits)

Orobanche legume strangler (one species parasitizes legumes (see also *rapum-genistae*)

orobus an old generic name for a leguminous plant

orontium an old generic name for a plant from the Orontium river, Syria

orophilus -a -um mountain-loving

ortho- correct-, upright-, straight-

Orthocarpus upright fruit

orubicus -a -um from Oruba Island, Caribbean

-osma, osmo- -scented, fragrant-

Osmanthus fragrant flower

Osmunda either for Osmund the waterman or for the Anglo-Saxon equivalent of Thor, god of thunder

ossifragus -a -um of broken bones (said to cause fractures in cattle when abundant in pastures)

osteo- bone-, bone-like-

ostruthius -a -um purplish

-osus -a -um -abundant, -large, -very much

-osyne, -otes -notably

ot-, oto- ear-, ear-like-

Otanthus ear flower (the shape of the corolla)

otites an old generic name, from Rupius, relating to ears

-otus -a -um -resembling, -having

ouletrichus -a -um with curly hair

ovali-, ovalis -is -e oval-, egg-shaped in outline

ovati-, ovatus -a -um egg-shaped (in the solid or in outline) with the broad end lowermost

ovifer -era -erum, oviger -era -erum bearing eggs (or egg-like structures)

ovinus -a -um of sheep

oxalis acid (a name used in Nicander refers to the taste)
oxy-, -oxys acid-, sharp-, -pointed
Oxyacantha sharp thorn (Theophrastus' name)
Oxycoccus acid berry
oxygonus -a -um sharp-angled, with sharp angles
oxylobus -a -um with sharp-pointed lobes
oxyphilus -a -um loving acid soils
Oxytropis sharp keel (the pointed keel petal)

pabularis -is -e of forage
pachy- thick-, stout-
Pachyphragma stout partition (the ribbed septum of the fruit)
Pachysandra thick stamens (the filaments)
pacificus -a -um of the western American seaboard
padus -a -um Theophrastus' name for St Lucie Cherry
Paeonia Theophrastus' name for Paeon, the physician, who, in
    mythology, was changed into a flower by Pluto
paganus -a -um of country areas, from the wild
palaestinus -a -um from Palestine
paleaceus -a -um covered with chaffy bracts, chaffy
palinuri of Palinura, Italy
Palisota for A. M. F. Palisot de Beauvois
Paliurus the ancient Greek for Christ-thorn
pallens pale
pallescens becoming pale, fading
palliatus -a -um cloaked, hooded
pallidus -a -um pale, greenish
palmaris -is -e of a hand's breadth, about 3 inches long
palmati-, palmatus -a -um with five or more veins arising from
    one point (usually of divided leaves), palmate (see Fig. 5(*a*))
palmensis -is -e from Las Palmas, Canary Isles
paludosus -a -um, paluster -tris -tre, palustris -is -e of boggy or
    marshy ground
pampinosus -a -um leafy
Panax healer of all (the ancient virtues of ginseng)
pancicii for Joseph Pančić
panduratus -a -um fiddle-shaped, pandurate
paniceus -a -um like millet grain
paniculatus -a -um with a branched-racemose inflorescence,
    paniculate (see Fig. 2(*c*))
Panicum the ancient Latin name
pannonicus -a -um from S.W. Hungary (Pannonia)
pannosus -a -um woolly, tattered, coarse

panormitanus -a -um from Palermo, Sicily

Papaver the Latin name for poppies, including the opium poppy

paphio- venus'-

papil-, papilio- butterfly-

papilliger -era -erum, papillosus -a -um having papillae or minute lobes on the surface, papillate

papyraceus -a -um with the texture of paper, papery

papyrifer -era -erum paper-bearing

Papyrus paper (the Greek name for the paper made from the Egyptian bulrush, *Cyperus papyrus*)

para- near-, besides-

paradisi, paradisiacus -a -um of parks, of paradise, of gardens

paradoxus -a -um strange, unexpected

paralias seaside (an ancient Greek name)

Parapholis irregular scales (the position of the glumes)

parasiticus -a -um parasitic (living on other plants and formerly including epiphytes)

parci- with few-

pardalianches, pardalianthes leopard-strangling (poisonous leopardsbane)

pardalinus -a -um, pardinus -a -um spotted or marked like a leopard

pardanthinus -a -um resembling *Belamcanda* (*Pardanthus*)

pari- equal-, paired-

parietarius -a -um of the wall (a name used by Pliny)

Paris equal (the regularity of its leaves and flower parts)

parmulatus -a -um with a small round shield

parnassi, parnassicus -a -um from Mt Parnassus, Greece

Parnassia Parnassus (the native home of Gramen Parnassi)

Paronychia beside nail (formerly used to treat whitlows)

Parthenium a Greek name for composites with white ray florets

Parthenocissus virgin ivy (Virginia creeper)

parthenus -a -um of the virgin, virginal

-partitus -a um -deeply divided, -parted

parvi-, parvus -a -um small

parvulus -a -um very small

Passiflora Passion flower (the signature of the numbers of parts in the flower to the events of the Passion)

pastinaca food, eatable

pastoralis -is -e growing in pastures, of shepherds

patagonicus -a -um from Patagonia, southern Argentina

patavinus -a -um from Padua, Italy

patellaris -is -e small dish-shaped

patens, patenti- spreading out from the stem, patent

patientia patience (corruption of patience dock, *Lapathum*)

patulus -a -um opened up, spreading

pauci-, paucus -a -um few

pauciflorus -a -um few-flowered

pauperculus -a -um poor

Pavonia, pavonianus -a -um for Don José Pavon

pavonicus -a -um peacock-blue

pavonius -a -um *Pavonia*-like, peacock-blue

pecten-veneris Venus' comb (a name used in Pliny)

pectinatus -a -um comb-like, pectinate

pectinifer -era -erum comb-bearing

pectoralis -is -e of the chest (used to treat coughs)

pedalis -is -e about a foot in length

pedatus -a -um palmate but with the lateral divisions sub-divided (see Fig. 5(*b*))

pedemontanus -a -um from Piedmont, N. Italy

pedicellulatus -a -um each flower borne upon its own stalk in the inflorescence

pedicularis -is -e of lice

pedifidus -a -um shaped like a bird's foot (of divided leaves)

pedil-, pedilo- shoe-, slipper-

Pedilanthus slipper flower

peduncularis -is -e, pedunculatus -a -um, pedunculosus -a -um with the inflorescence supported on a distinct stalk, pedunculate

pel- through-

Pelargonium stork (Greek name compares the fruit-shape of florists' Geraniums with a stork's head)

pellucidus -a -um clear, transparent, pellucid

pelorius -a -um monstrous, peloric (as with radial forms of normally bilateral flowers)

peltatus -a -um stalked from the surface (not the edge), peltate (see Fig. 5(*d*))

pelviformis -is -e shallowly cupped, shaped like a shallow bowl

pendens, penduli-, pendulinus -a -um, pendulus -a -um hanging down, drooping

penicillatus -a -um, penicillius -a -um covered with tufts of hair, brush-like

peninsularis -is -e living on a peninsula

pennatus -a -um, penni-, penniger -era -erum arranged like the barbs of a feather, feathered

pennivenius -a -um pinnately veined

pensilis -is -e hanging down

pensylvanicus -a -um from Pennsylvania, U.S.A

pent-, penta- five-

Pentaglottis five tongues (the scales in the throat of the corolla)

Peperomia pepper-like, resembling *Piper*

peplis Dioscorides' name for a Mediterranean coastal spurge

peplodes peplus-like

peplus Dioscorides' name for the northern equivalent of *peplis*

Pepo sun-cooked (ripens to become edible)

per- through-, beyond-, extra-, very-

peramoenus -a -um very pleasing, very beautiful

percussus -a -um actually or appearing to be perforated

peregrinus -a -um strange, foreign, exotic

perennans, perennis -is -e through the years, continuing, perennial

perfoliatus -a -um, perfosus -a -um the stem appearing to pass through completely embracing leaves

perforatus -a -um pierced or apparently pierced with round holes

peri- around-

periclymenum Dioscorides' name for a twining plant

Periploca twine around

permixtus -a -um confusing

perpropinquus -a -um very closely related

perpusillus -a -um very small

persicarius -a -um resembling peach (the leaves)

persici- peach-

persicus -a -um from Persia, Persian

persistens persistent

perspicuus -a -um transparent

persutus -a -um with slits or holes

pertusus -a -um pierced through, perforated

perulatus -a -um with conspicuous scales (e.g. on buds)

peruvianus -a -um from Peru, Peruvian

-pes -foot, -stalk

pes-caprae goat's foot

pes-tigridis tiger's foot

Petasites Discorides' name (use of leaves as hats)

petecticalis -is -e blemished with spots

petiolaris -is -e, petiolatus -a -um having a petiole, not sessile, petiolate

petiolosus -a -um with a well-developed petiole

petr-, petro-, petraeus -a -um stony, rocky, of rocky places

Petroselinum Dioscorides' name for parsley

Peucedanum a name used by Theophrastus

Petunia from the Brazilian 'petun' tobacco
phaeno- shining
phaeo-, phaeus -a -um dusky-brown, dark
Phalaris Dioscorides' name for a grass
phanero- conspicuous-
Phaseolus the old Latin name for a kind of bean
Phegopteris oak fern
phello- cork-, corky-
phil-, philo-, -philus -a -um -loving, liking, -fond of
philadelphicus -a -um from Philadephia
Philadelphus brotherly love
Philonotis moisture lover
phlebanthus -a -um with veined flowers
phleioides rush-like, resembling *Phleum*
Phleum copious (Greek name for a kind of dense-headed rush)
-phloebius -a -um -veined
phlogi- flame-, *Phlox*-like-
Phlomis flame (the hairy leaves were used as lamp-wicks)
phoeniceus -a -um red-purple, from Tyre and Sidon (Phoenicia)
Phoenix Greek name for the date palm
-phorus -a -um -bearing, -carrying
Phragmites hedge dweller (common habitat)
phrygius -a -um from Phrygia, Asia Minor
phu foul-smelling
Phuopsis valerian-like, resembling *Valeriana phu*
Phylica leafy (copious foliage)
phylicifolius -a -um with leaves like those of *Phylica*
phyll- leaf-
Phyllitus a name used by Dioscorides
Phyllodoce the name of a sea nymph
phyllomaniacus -a -um excessively leafy, a riot of foliage
-phyllus -a -um -leaved
physa- bladder-
Physalis bladder (the enlarged calyx in fruit)
physo- inflated-, bellows-
Phyteuma a name used by Dioscorides for rampion
phyto- plant-
Phytolacca plant dye (the staining sap)
Picea pitch (the resin)
piceus -a -um black, blackening
picridis -is -e of *Picris* (Theophrastus' name for a potherb)
picturatus -a -um variegated
pictus -a -um painted, brightly marked

124

pileatus -a -um capped, having a cap

pilifer -era -erum hairy with short soft hairs

pilo-, pilosus -a -um felted with long soft hair

Pilularia small balls (the sporocarps)

pilularis -is -e, pilulifer -era -erum bearing small balls or globular structures

pinaster Pliny's name for *Pinus sylvestris*

pinetorus -a -um of pine woods

pineus -a -um of pines, resembling a pine

pini- pine-like-

pingui- fat-

Pinguicula fat (the leaves)

pinnati-, pinnatus -a -um set in two opposite ranks, pinnate (of the leaflets of a compound leaf or of leaf veins) (see Fig. 5(*c*))

Pinus the Latin name for a pine

Piper from the Indian name for pepper

piperitius -a -um peppery (the taste)

piri- pear-

Pirola small pear (similarity of the leaves)

Pirus the Latin name for the pear tree

pisi-, piso- pea-, pea-like-

pisifer -era -erum bearing peas

Pistia watery (the habitat of the water-lettuce)

Pisum the Latin name for the pea

pithece-, pitheco- ape-, monkey-

Pittosporum pitch seed (the resinous coating of the seed)

placatus -a -um quiet, calm, gentle

plagio- oblique

planetus -a -um wandering

plani-, planus -a -um flat

plantagineus -a -um plantain-like, ribwort-like

Plantago foot sole (the way plantain leaves lie flat on the ground)

Platanthera broad anthers

plat-, platy- flat-, broad-

Platanus flat leaf (the Greek name for a plane tree)

plebio-, plebius -a -um common

plecto-, plectus -a -um pleated

plectro-, plectrus -a -um spur, spurred

pleio-, pleo- many-, more-, full-, large-, thick-, several-

pleni-, plenus -a -um full, double

pleniflorus -a -um double-flowered

plesio- near to-, close by-

pleuro- ribs-, edge-, side-, of the veins-

plicatus -a -um, plici-, ploco- pleated, folded lengthwise

plumarius -a -um, plumatus -a -um feathery, plumose

Plumbago leaden (Pliny's name refers to the flower colour)

plumbeus -a -um lead-coloured

plumosus -a -um feathery

pluri- several-, many-

pluvialis -is -e, pluviatilis -is -e of rainy places

pneumonanthis -is -e lung flower (former use for respiratory disorders, of Gentian)

Poa the Greek name for a grass

pocophorus -a -um fleece-bearing

podagrarius -a -um, podagricus -a -um snare, of gout (used to treat gout)

-podioides -foot-like

-podius -a -um, podo-, -podus -a -um foot, stalk

poecilo- variable-, variegated-, spotted-

poetarum, poeticus -a -um of poets (Greek gardens included games areas and theatres)

-pogon -haired, -bearded

poikilo- variable-, variegated-

poissonii for M. Poisson

polaris -is -e of the North Pole

Polemonium Pliny's name after King Polemon of Pontus

polifolius -a -um grey-coloured leaves, *Teucrium*-leaved

polio- grey-

politus -a -um elegant, polished

pollicaris -is -e as long as the end joint of the thumb, about an inch long

polonicus -a -um from Poland, Polish

poly- many-

polyanthemos, polyanthus -a -um many-flowered

Polycarpon many fruits (the name given by Hippocrates)

polyedrus -a -um many-sided

Polygala much milk (Dioscorides' name refers to the improved lactation in cattle fed on it)

polygamus -a -um the flowers having various combinations of the sexual structures

Polygonatum many knees (the rhizome structure)

Polygonum many knees (the swollen stem nodes)

polygyrus -a -um twining

Polystichum many rows (arrangement of the sori on the fronds)

pomeridianus -a -um of the afternoon (afternoon flowering) *pomi-, pomaceus -a -um* apple-like

pomifer, -era -erum apple-bearing

pomponius -a -um of great splendour, pompous

ponderosus -a -um heavy, large

Pontederia for Guilio Pontedera, former Professor of Botany at Padua

ponticus -a -um of the Black Sea's southern area

populifolius -a -um poplar-leaved

populneus -a -um poplar-like, related to *Populus*

Populus the ancient name for poplar

porcinus -a -um of pigs

porophilus -a -um loving soft stony ground

porophyllus -a -um having holes in the leaves

porosus -a -um with holes or pores

porphyreus -a -um, porphyrion warm-reddish-coloured

porri- leek-, leek-like-

porrifolius -a -um leek-leaved

porrigens spreading

porrigentiformis -is -e porrigens-like (with leaf-margin teeth pointing outwards and forwards)

porum a Latin name used for various onions

portensis -is -e from Oporto

portlandicus -a -um from the Portland area

portulus -a -um somewhat porous

post- behind-, after-, later-

potamo- watercourse, of watercourses-

Potamogeton watercourse neighbour

potamophilus -a -um river-loving

potatorum of drinkers (used for fermentation)

Potentilla quite powerful (as a medicinal herb)

Poterium drinking cup (the name was earlier used for some other plant)

-pous -foot, -stalk

prae- before-, in front-

praealtus -a -um very tall or high

praecox earlier than most of its genus, forward, very early developing

praemorsus -a -um as if nibbled at the tip

praeruptorum of rough places (living on screes, etc.)

praestans excelling, distinguished

praetermissus -a -um overlooked, omitted

praetextus -a -um bordered

praevernus -a -um before spring, early

prasinus -a -um, *prasus -a -um* leek-like, leek-green

pratensis -is -e of meadows

pratericolus -a -um, *praticolus -a -um* of meadows, living in grassy places

pravissimus -a -um very crooked

precatorius -a um relating to prayer (rosary beads)

prenans drooping

Prenanthes drooping flower (the nodding flowers)

preptus -a -um eminent

Primula little first one

primulinus -a -um *Primula*-like, primrose-coloured

primuloides resembling *Primula*

princeps most distinguished

priono- serrated-, saw-toothed-

prismati- prism-, prism-like-

pro- forwards-, before-, for-, instead of-

proboscoideus -a -um snout-like, trunk-like

procerus -a -um very tall

procumbens prostrate, lying flat on the ground, creeping forwards

procurrens spreading below ground, running forwards

prodigiosus -a -um wonderful, marvellous

productus -a -um stretched out, produced

profusus -a -um very abundant, profuse

prolifer -era -erum producing offsets or young plantlets or bunched growth, proliferous

prolificus -a -um very beautiful, very fruitful

pronus -a -um with a forward tilt, lying flat

propaguliferus -a -um prolific, multiplying by vegetative propagules

propensus -a -um hanging down

propinquus -a -um closely allied, of near relationship, related

pros- near-, in addition-, also-

proso-, *prostho-* towards-, to the front-, before-

prostratus -a -um lying flat but not rooting

Protea for Proteus, the versatile sea god of mythology

protrusus -a -um protruding

provincialis -is -e from Provence, France

pruhonicus -a -um from Průhonice, Czechoslovakia

pruinatus -a -um, *pruinosus -a -um* with a glistening surface as though frosted over

Prunella from a German name in reference to its use in treating quinsy

pruni- plum-
Prunus the Latin name for a plum tree
pruriens irritant, stinging
pseud-, pseudo- sham-, false-
Pseuderanthemum false eranthemum
Psidium a Greek name, formerly for the pomegranate
psilo- bare-, smooth-
psittacinus -a -um parrot-like (the colouration)
psittacorum of parrots
psycodes butterfly-like
psyllius -a -um flea-like (the appearance of the seeds)
ptarmicus -a -um causing sneezes
ptera-, ptero-, -pteris, -pterus -a -um -winged
Pteridium small fern
Pteris the Greek name for a Fern (feathery)
Pterocarya wing nut (the winged fruits of most)
ptilo- feathery-
ptycho- folded-
pubens, pubescens, pubiger -era -erum softly hairy, covered with
    down, downy
puddum from a Hindi name for a cherry
puderosus -a -um very bashful
pudicus -a -um modest, bashful, retiring
puellii for Timothée Puel, French botanist
pugioniformis -is -e dagger-shaped
pulchellus -a -um pretty
pulcher -ra -rum beautiful, fair
pulcherrimus -a -um very beautiful
Pulegium flea repeller (the Latin name)
Pulicaria fleabane
pulicaris -is -e of fleas (e.g. the shape of the fruits)
pullus -a -um raven-black, almost dead-black
Pulmonaria lung wort (the signature of the spotted leaves as
    indicative of efficacy in treating respiratory disorders)
Pulsatilla pulsator, quiverer (movement of flowers in wind)
pulverulentus -a -um covered with powder, powdery
pulvinatus -a -um cushion-shaped, cushion-like
pumilus -a -um low, small, dwarf
punctati-, puncti-, punctatus -a -um with a pock-marked surface,
    spotted
puncticulatus -a -um minutely dotted
pungens ending in a sharp point, pricking
puniceus -a -um crimson, carmine-red

purgans, purgus -a -um purgative

purpurascens becoming purple

purpuratus -a -um purplish, empurpled

purpureus -a -um reddish-purple

purpusii for either of the brothers J. A. and C. A. Purpus

-pus -foot

pusillus -a -um very small, minute, insignificant

pustulatus -a -um covered with blisters, pustuled

puteorus -a -um of the pits

pynco- compact-, dense-, densely-

pygameus -a -um dwarf

Pyracantha fire thorn (persistent irritation from thorn pricks, and the appearance in fruit)

pyramidalis -is -e, pyramidatus -a -um conical, pyramidal

pyrenaeus -a -um, pyrenaicus -a -um from the Pyrenees

Pyrethrum fire (medicinal use in treating fevers)

pyri- pear-

Pyrus the ancient Latin name for a pear tree

pyxidatus -a -um with a lid, box-like (e.g. some stamens)

quadrangularis -is -e, quadrangulatus -a -um with four angles, quadrangular

quadratus -a -um into four, in fours

quadri four-

quadrifidus -a -um divided into four, cut into four

Quamoclit the Greek name formerly for a bean

quarternallus -a -um four-partite, with four divisions

querci-, quercinus -a -um oak, oak-like

Quercus the old Latin name for an oak

-quetrus -a -um -acutely-angled, -angled

quin- five-

quinatus -a -um five partite, with five divisions, in fives

quinque- five-

quinquevulnerus -a -um five-wounded, with five marks (e.g. on corolla)

Quisqualis Who? What? (from a Malay name 'udani' which Rumphius transliterated as Dutch 'hoedanig' meaning 'how, what')

quitensis -is -e from Quito, Ecuador

racemi-, racemosus -a -um with flowers arranged in a raceme (see Fig. 2(b))

radians, radiatus -a -um radiating outwards

radicans with rooting stems

radicatus -a -um with a large or conspicuous root

radiosus -a -um with numerous roots

Radiola radiating (the branches)

radiosus -a -um having many rays

radula a rasp

ragusinus -a -um from Dubrovnik (Ragusa), Yugoslavia

ramentaceus -a -um covered with scales

rami- branches-, of branches-, branching-

ramosissimus -a -um greatly branched

ramosus -a -um branched

ramulosus -a -um twiggy

Ranunculus little frog (the amphibious habit of many)

rapa an old Latin name for a turnip

rapaceus -a -um of turnips, *Rapa*-like

raphani- radish-, radish-like-

Raphanus the Latin name for a radish

rapum-genistae rape of broom (a parasite of *Sarothamnus*)

rapunculoides resembling *Rapunculus*, rampion-like

rari-, rarus -a -um uncommon, scattered

Ravenala from the Madagascan name for the Traveller's tree

ravus -a -um tawny-grey-coloured

re- back-, again-, against-

reclinatus -a -um drooping to the ground, reflexed, reclining

rectus -a -um straight, upright, erect

recurvus -a -um curved backwards

recutitus -a um circumcised (the appearance of the flower heads
    with the rays reflexed)

redivivus -a -um coming back to life, renewed (perennial habit or
    reviving after drought)

reductus -a -um reduced, drawn back

reflexus -a -um bent back upon itself, reflexed

refractus -a -um abruptly bent, splitting open

regalis -is -e royal, outstanding, kingly

reginae of the queen

regius -a -um royal, splendid, princely

religiosus -a -um sacred, of religious rites

remotus -a -um scattered (e.g. the flowers on the stalk)

reniformis -is -e kidney-shaped, reniform

repandus -a -um with a slightly wavy margin

repens creeping (stoloniferous)

replicatus -a -um double-pleated, doubled down

reptans creeping

131

Reseda Healer (the name in Pliny refers to its use in treating bruises)

resinifer -era -erum, resinosus -a -um producing resin, resinous

resupinatus -a -um inverted (e.g. those orchids with twisted ovaries)

reticulatus -a -um netted, conspicuously net-veined

retro-, retroflexus -a -um, retrofractus -a -um, retrorsus -a -um directed backwards and downwards

retusus -a -um shallowly notched at the tip (e.g. leaves, see Fig. 7(*f*))

revolutus -a -um rolled back, rolled out and under (e.g. leaf margin)

rex king

rhabdotus -a -um striped

rhaeticus -a -um from the Rhaetian Alps in the Swiss-Austrian border

-rhagius -a -um -torn, -rent

Rhamnus an ancient name for various prickly shrubs

rhaponticus -a -um from the Black Sea area

Rheum from a Parisian name for rhubarb

Rhinanthus nose flower

rhiz-, rhizo-, -rhizus -a -um root-, -rooted

rhodensis -is -e, rhodius -a -um from the Aegean Island of Rhodes

rhodo- rose-, rosy-, red-

Rhododendron rose tree (a name formerly used for an oleander)

rhodopensis -is -e from Rhodope Mountain, Bulgaria

rhoeas the old generic name of the field poppy

Rhoicissus sumach ivy (resemblance to *Rhus*)

rhombicus -a -um, rhomboidalis -is -e, rhomboidosus -a -um diamond-shaped, rhombic

rhopalo- club-, cudgel-

Rhynchelytron (um) beak flower (the shape of the spikelets)

Rhynchosia beak (the shape of the keel petals)

rhytido- wrinkled-

rhytidophyllus -a -um with wrinkled leaves

Ribes from the Persian for acid-tasting

Ricinus tick (the appearance of the seed)

rigens, rigidus -a -um stiff, rigid

rimosus -a -um with a cracked surface, furrowed

ringens gaping, with a two-lipped mouth

riparius -a -um of the banks of streams and rivers

ritro a southern European name for *Echinops ritro*

rivalis -is -e of brooksides and streamsides

riviniana for A. Q. Rivinus, formerly Professor of Botany at
  Leipzig
rivularis -is -e of the waterside, of streamsides
robbiae for Mrs Robb, who introduced *Euphorbia robbiae* from
  Turkey
robertianus -a -um of Robert (which Robert is uncertain)
robur strong, hard, oak timber
robustus -a -um strong-growing
romanus -a -um from Rome
Romulea for Romulus, founder of Rome
Rorippa from an old Saxon name
rorulentus -a -um dewy
Rosa the Latin name for various roses
rosaceus -a -um looking or coloured like a rose
rosae-, rosi-, roseus -a -um rose-coloured, rose-like
Rosmarinus seaside dew
rostellatus -a -um with a small beak, beaked
rostratus -a -um with a long straight hard point, beaked
rosularis -is -e with leaf-rosettes
rotatus -a -um flat and circular, wheel-shaped
rotundi-, rotundus -a -um rounded in outline or at the apex,
  spherical
-rrhizus -a -um -rooted
rubellus -a -um reddish
rubens blushed with red
ruber, rubra, rubrum, rubri-, rubro- red
rubescens, rubidus -a -um turning red, reddish
Rubia a name in Pliny for madder
rubicundus -a -um ruddy, reddened
rubiginosus -a -um, rubrus -a -um rusty red
Rubus the Latin name for brambles
ruderalis -is -e of waste places, of rubbish tips
rudis -is -e wild, rough, untilled
rufescens reddish, turning red
rufinus -a -um red
rufus -a -um pale brownish or reddish-brown
rugosus -a -um wrinkled, rugose (e.g. leaf surfaces)
rugulosus -a -um somewhat wrinkled
Rumex a name in Pliny for sorrel
ruminatus -a -um thoroughly mingled
runcinatus -a -um saw-toothed, sharply cut (e.g. leaf margins)
rupester -tris -tre, rupestris -is -e, rupicola of rock, of rocky places
rupifragus -a -um growing in rock crevices, rock cracking

133

rupri- of rocks or rocky places

ruralis -is -e of country places, rural

rurivagus -a -um of country roads

rusci- Ruscus-like-, resembling box holly or butcher's broom

Ruscus an old name for a prickly plant

russatus -a -um reddened, russet

russotinctus -a -um red-tinged

rusticanus -a -um, rusticus -a -um of wild places, of the country-side, rustic

Ruta the Latin name for rue

ruta-baga from a Swedish name

ruta-muraria rue of the wall

ruthenicus ₌a -um from Ruthenia, Russia

rutilans, rutilus -a -um deep bright glowing red, orange-red

rytido- wrinkled-

sabatius -a -um from Savona, N.W. Italy

sabaudus -a -um from Savoy, S.E. France

sabdariffa from a West Indian name

sabrinae of the River Severn

sabulicolus -a -um sand-dweller, living in sandy places

sabulosus -a -um of sandy ground, full of sand

saccatus -a -um bag-shaped, saccate

saccharatus -a -um with a scattered white coating, sugared, sweet-tasting

sacchariferus -a -um sugar-producing, bearing sugar

saccharinus -a -um, saccharus -a -um sweet, sugary

saccifer -era -erum having a hollowed part, bag-bearing

sachalinensis -is -e from Sakhalin Island, eastern U.S.S.R.

sacrorum of sacred places, of temples (former ritual use)

saepius -a -um of hedges

Sagina fodder (the virtue of a former included species, spurrey)

sagittalis -is -e, sagittatus -a -um arrow-shaped, sagittate (see Fig. 6(*c*))

Sagittaria arrowhead (the shape of the leaf-blades)

salebrosus -a -um rough

salicarius -a -um, salice-, salici-, salicinus -a -um willow-like, willow-

Salicornia salt horn

salignus -a -um of willow-like appearance, resembling *Salix*

salinus -a -um of salt-marshes, halophytic

salisburgensis -is -e from Salzburg, Austria

Salix the Latin name for willows

salpi- trumpet-

Salpiglossis trumpet tongue (the shape of the style)
salsus -a -um of saline soils, salted
Salsola salt (the taste and the habitat)
salsuginosus -a -um living in saline soils
saltatorius -a -um dancing
saltitans jumping
saltuum woodland, pasture
salutaris -is -e healing
Salvia healer (the old Latin name for sage)
salvii- sage-like, resembling *Salvia-*
salviodorus -a -um sage-scented
saman, Samanea from a south American name
sambucinus -a -um elder-like, resembling *Sambucus*
Sambucus the Greek name for the elder tree
saminus -a -um from the Isle of Samos, Greece
Samolus from a Celtic Druidic name
sanctus -a -um holy
sanguinalis -is -e, sanguineus -a -um, sanguineolentus -a -um
    blood-red
Sanguisorba blood-absorber (has styptic property)
Sanicula healer
Santolina holy flax
sapidus -a -um pleasant-tasted, flavoursome, savoury
sapientius -a -um of the wise, of man (implies superiority,
    compare with *troglodytarum*)
Sapindus Indian soap (from its use)
saponaceus -a -um, saponarius -a -um lather-forming, soapy
sapota a South American name for *Sapodilla*, chicle tree
saracenicus -a -um, sarracenicus -a -um of the Saracens
sarachoides resembling *Saracha*
sarc-, sarco- fleshy-
sarcodes flesh-like
sardensis -is -e from Sardis, Smyrna
sardosus -a -um, sardous -a -um from Sardinia, Sardinian
sarmaticus -a -um from Sarmatia, on the Russo-Polish borders
sarmentaceus -a -um, sarmentosus -a -um with long slender
    stolons or runners
sarniensis -is -e from Guernsey (Sarnia), Channel Isles
saro- broom-like
Sarothamnus broom shrub (brush bush)
Sasa the Japanese name for certain bamboos
sassafras from the Spanish name for saxifrage
sativus -a -um not wild, planted, cultivated, sown

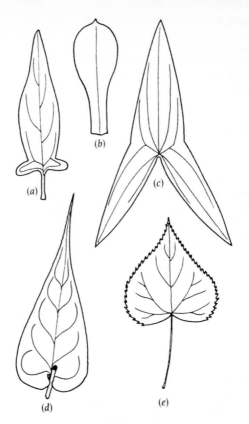

Fig. 6. More leaf shapes which provide specific epithets:
(a) hastate (e.g. *Scutterlaria hastifolia* L.), with auricled leaf-base;
(b) spathulate (e.g. *Sedum spathulifolium* Hook.); (c) sagittate (e.g.
Sagittaria sagittifolia L.), with pointed and divergent auricles;
(d) amplexicaul (e.g. *Polygonum amplexicaule* D. Don), with the basal
lobes of the leaf clasping the stem; (e) cordate (e.g. *Tilia cordata* Mill.),
heart-shaped.

Satureia, Satureja the Latin name for a culinary herb

sauro- lizard-, lizard-like-

Saussurea for the Swiss H. B. de Saussure

saxatilis -is -e of rocky places, rock-liking

saxicola, saxosus -a -um rock-dweller

Saxifraga stone breaker (lives in rock cracks and had a medicinal use for gall stones)

scaber -ra -rum, scabri- scurfy, rough

scaberulus -a -um, scabriusculus -a -um roughish, somewhat rough

Scabiosa scabies (former medicinal treatment for)

scabrosus -a -um rather rough

scalaris -is -e ladder-like

scandens climbing

scandicus -a -um from Scania

Scandix ancient name for the shepherd's needle

scaphi-, scapho-, scaphy- boat-shaped-, bowl-shaped-

scapi- clear-stemmed-

scapiger -era -erum scape-bearing

scaposus -a -um with scapes or leafless flowering stems

scariolus -a -um (see *serriolus*)

scariosus -a -um shrivelled, thin

scarlatinus -a -um bright-red

sceleratus -a -um of vile places, vicious, wicked (causes ulceration)

sceptrum sceptre

schinseng from the Chinese name

schis-, schiz-, schismo-, schizo- divided-, cut-

Schizanthus divided flower (the lobes of the corrola are subdivided in the poor man's orchid)

schoeno- rush-like-, resembling *Schoenus-*

schoenoprasus -a -um rush-like leek (the leaves)

Schoenus the old name for rush-like plants

sciadi-, sciado- shade-, canopy-, shaded-, umbelled-

sciaphilus -a -um shade-loving

Scilla the ancient Greek name for a squill

Scindapsus the ancient Greek name for an ivy-like plant

scintillans sparkling, gleaming

Scirpus the old name for a rush-like plant

scitulus -a -um neat, pretty

scitus -a -um fine

sciuroides squirrel-tail-like

sclareus -a -um clear (from a name for a *Salvia* used for eye lotions, clary)

Scleranthus hard flower (the texture of the perianth)

sclero- hard-

Scolopendrium Dioscorides' name for the hart's tongue fern,
  (compares the numerous sori to a millipede's legs)

scoparius -a -um broom-like (use for twig-brushes)

scopulinus -a -um twiggy

scopulorum of cliffs and rock faces

scorbiculatus -a -um with surface depressions or grooves

Scordium Dioscorides' name for a plant with the smell of garlic

scorodonius -a -um an old generic name for garlic

scorodoprasus -a -um garlic-like leek (has intermediate features)

scorpioides curved like a scorpion's tail (see Fig. 3)

scorteus -a -um leathery

Scorzonera derivation uncertain but generally thought to refer to
  use as an antifebrile in snakebite

scoticus -a -um from Scotland, Scottish

scotinus -a -um dusky, dark

scottianus -a -um for Munro B. Scott

scriptus -a -um marked with lines which suggest writing

Scrophularia scrophula (the glands on the corolla)

sculptus -a -um carved

scutatus -a -um like a small round shield

Scutellaria dish (the depression of the fruiting calyx)

scutellatus -a -um shield-shaped, platter-like

scypho- cup-, beaker-

se- apart-, without-, out-

sebaceus -a -um, sebifer -era -erum producing wax, waxy

sebosus -a -um full of wax

Secale the Latin name for a grain like rye

secalinus -a -um rye-like, resembling *Secale*

sechellarum from the Seychelles, Indian Ocean

seclusus -a -um hidden, secluded

secundus -a -um one-sided (as when flowers are all to one side of
  an inflorescence)

securiger -era -erum axe-bearing (the shape of some organ)

Securinega axe-refuser (the hardness of the timber)

sedi-, sedioides stonecrop-like, resembling *Sedum*

Sedum a name in Pliny refers to the plants 'sitting' on rocks etc.
  in the case of cushion species

segetalis -is -e, segetus -a -um of cornfields

Selaginella a diminutive of selago (see below)

selaginoides clubmoss-like, resembling *Selaginella*

selago an old generic name for *Lycopodium* which is applied in
Pliny to a ceremonial plant

seleni- moon-

Selinum the ancient name for a celery-like plant

selloi, sellovianus -a -um, sellowii, seloanus -a -um for Friedrich
Sellow (Sello), German botanist

semi- half-

semper- ever-, always-

sempervirens always green

sempervivoides houseleek-like, resembling *Sempervivum*

Sempervivum never-die, always alive

Senecio old man (the name in Pliny refers to the grey hairiness
as soon as fruiting commences)

senescens turning hoary with whitish hairs

senilis -is -e aged, grey-haired

sensibilis -is -e, sensitivus -a -um sensitive to touch

senticosus -a -um thorny, full of thorns

sepiarius -a -um, sepius -a -um of hedges, growing in hedges

sepincolus -a -um inhabiting hedges, hedge-dweller

sept- seven-

septentrionalis -is -e northern, of the north

sepulchralis -is -e of tombs of graveyards

serapias an ancient name for an orchid

seri-, serici-, sericans, sericeus -a -um silky-hair (sometimes
implying Chinese)

sericifer -era -erum, sericofer -era -erum silk-bearing

-seris -potherb

serotinus -a -um of late season

serpens creeping, serpentine

serpyllifolius -a -um thyme-leaved

serra- saw-, saw-like-, serrate-

Serrafalcus for the Duke of Serrafalco, archaeologist

Serratula saw tooth (the name in Pliny for betony)

serratus -a -um edged with forward-pointing teeth, serrate (see
Fig. 4(*c*))

serriolus -a -um in ranks, of salad (one form of an old name for
chicory)

serrulatus -a -um edged with small teeth, finely serrate

Sesamum from the Semitic name

Seseli the ancient Greek name

Sesleria for Leonardo Seslero of Venice

sesqui- one and a half-

sesquipedalis -is -e about 18 inches long, the length of a foot and a half

sessili-, sessilis -is -e attached without a distinct stalk, sessile

setaceus -a -um, seti- bristly, with bristles or stiff hairs

setifer -era -erum, setiger -era -erum bearing bristles

setosus -a -um covered with bristles or stiff hairs

setulosus -a -um slightly bristly

sex- six-

shallon from a Chinook Indian name

siameus -a -um from Thailand (Siam)

sibiricus -a -um from Siberia, Siberian

siccus -a -um dry

siculi- dagger-shaped-

siculus -a -um from Sicily, Sicilian

Sida from a Greek name for a water-plant

sidereus -a -um iron-hard

Sideritis the Greek name for plants used to dress wounds caused by iron weapons

Sideroxylon iron wood (the hard timber of the miraculous berry)

signatus -a -um well-marked, signed

silaus an old generic name, in Pliny used for pepper saxifrage

Silene Theophrastus' name for *Viscaria*, another catchfly

siliceus -a -um growing on sand

siliculosus -a -um, siliquosus -a -um having elongate pods

siliquastrum the old Latin name for a pod-bearing tree

silvaticus -a -um, silvester -tris -tre of woods, of woodlands, wild

Silybum Dioscorides' name for a thistle-like plant

simensis -is -e from Arabia, Arabian

similis -is -e resembling other species

simius -a -um ape, monkey (flower shape or implying inferiority)

simplex, simplici- undivided, simple

simulans, simulatus -a -um resembling, imitating

Sinapis the name used by Theophrastus for mustard

sinensis -is -e from China, Chinese

singularis -is -e unusual, singular

sinicus -a -um, sino- Chinese

sinuatus -a -um, sinuosus -a -um waved, with a wavy margin, sinuate (see Fig. 4(*e*))

siphiliticus -a -um used to treat the disease (*Lobelia siphilitica*)

sipho-, -siphon tubular-, -pipe, -tube

sisalanus -a -um from Sisal, Yucatan, Mexico

sisarus -a -um Dioscorides' name for a plant with an edible root

Sison a name used by Dioscorides

Sisymbrium ancient Greek name for various plants
Sisyrinchium Theophrastus' name for an iris
sitchensis -is -e from Sitka Island, Alaska
Sium an old Greek name for water plants
Skimmia from a Japanese name
Smyrnium myrrh-like (the fragrance)
sobolifer -era -erum producing vigorous shoots from the stem at
    ground level
socialis -is -e growing in colonies
sodomeus -a -um from the Dead Sea area (Sodom)
solani- potato-, *Solanum-*
Solanum a name in Pliny
solaris -is -e of the sun, of sunny habitats
Soldanella coin-shaped (the leaves)
solen-, soleno- box-, tube-
Solidago uniter (use as a healing medicine)
solidus -a -um a coin, solid, dense
solstitialis -is -e of mid-summer (flowering time)
somnians asleep, sleeping
somnifer -era -erum sleep-bearing, sleep-inducing
Sonchus the Greek for a thistle
sophia wisdom (the use of flixweed in healing)
soporificus -a -um sleep-bringing, soporific
Sorbus the Latin name for the service tree
sordidus -a -um neglected, dirty-looking
sororius -a -um very closely related, sisterly
spadiceus -a -um chestnut-brown, date-coloured
Sparganium Dioscorides' name for bur-reed
sparsi-, sparsus -a -um scattered
Spartina old name for various plants used to make ropes
Spartium binding or broom (former uses for sweeping and
    binding)
spathi-, spatho- spathe- (as in the arums)
spathulatus -a -um shaped like a spoon, spathulate (see Fig. 6(*b*))
speciosus -a -um showy, handsome
spectabilis -is -e admirable, spectacular
specularia, speculus -a -um shining, mirror-like
speluncarus -a -um of caves
speluncatus -a -um shining
-spermus -a -um -seeded
sphacelatus -a -um necrotic, gangrened
sphaer-, sphaero- globular-
sphaerocephalus -a -um round-headed

sphegodes resembling wasps (flower shape)

spheno- wedge-

sphondilius -a -um rounded

spicant tufted (spikenard, spike, ear)

spicatus -a -um, *spicati-*, *spicifer -era -erum*, *spicus -a -um* with an elongate inflorescence of sessile flowers, spiked, spicate (see Fig. 2(*a*))

spica-venti ear of the wind, tuft of the wind

spina-christi Christ's thorn

spinescens, *spinifer -era -erum*, *spinifex*, *spinosus -a -um* spiny, with spines

spinulifer -era -erum, *spinulosus -a -um* with small spines

Spiraea Theophratus' name for a plant used for making garlands

spiralis -is -e twisted, spiral

Spiranthes twisted (the inflorescence)

splendens, *splendidus -a -um* gleaming, striking

spongiosus -a -um spongy

sponhemicus -a -um from Sponheim, Rhine

-sporus -a -um -seed, -seeded

spretus -a -um spurned

spumarius -a -um frothing

spurius -a -um false

squalens, *squalidus -a -um* untidy, dingy, squalid

squamarius -a -um scale-clad, covered with scales

squamatus -a -um scaly, squamate

squamosus -a -um with scale-like leaves, full of scales

squarrosus -a -um rough (as when small overlapping leaves have protruding tips), spreading in all directions

stachy- spike-like, resembling *Stachys*

Stachys spike (the Greek name for several deadnettles)

stachyon, *-stachys*, *-stachyus -a -um* -spikeleted, panicled

Stachytarpheta thick spike

stagnalis -is -e of pools

stagninus -a -um of swampy or boggy ground

stamineus -a -um with prominent stamens

stans upright, erect

Staphylea cluster (a name in Pliny refers to the flower)

-staphylos -bunch (of grapes)

Statice astringent (Dioscorides' name for *Limonium* of gardeners)

stauro- cross-shaped, crosswise-, cruciform-

steiro- barren-

stellaris -is -e, stellatus -a -um star-like, with spreading rays, stellate

Stellaria star (the appearance of stitchwort flowers)

stelliger -era -erum star-bearing

-stemon -stamened

sten-, steno- narrow-

stephan-, stephano- crown-

Stephanotis crown (the auricled staminal crown). Also used by the Greeks for other plants used for making chaplets or crowns

stepposus -a -um of the Steppes

sterilis -is -e seed-sterile, with barren fruit

-stichus -a -um -ranked, -rowed

stict-, sticto- punctured-, spotted-

stictocarpus -a -um spotted fruit

stigmaticus -a -um spotted, dotted, marked

stigmosus -a -um spotted

Stipa tow (Greek use of the feathery inflorescences, like hemp, for caulking and plugging)

stipellatus -a -um with stipels (in addition to stipules)

stipitatus -a -um with a stipe or stalk

stipulaceus -a -um, stipularis -is -e, stipulatus -a -um, stipulosus -a -um with conspicuous stipules

stolonifer -era -erum spreading by stolons, with creeping stems which root at the nodes

stragulus -a -um carpeting, mat-forming

stramineus -a -um straw-coloured

stramonius -a -um spiky-fruited (used as a name by Theophrastus for the thorn-apple)

strangulatus -a -um constricted, strangled

Stratiotes solider (Dioscorides' name for an Egyptian water plant)

strepens rustling, rattling

strept-, strepto- twisted-, coiled-

striatellus -a -um, striatulus -a -um, striatus -a -um marked with parallel lines grooves or ridges

stricti-, stricto-, strictus -a -um straight, erect

strigilosus -a -um with short appressed bristles

strigosus -a -um with rigid hairs or bristles, strigose

strigulosus -a -um somewhat strigose

striolatus -a -um faintly striped

strobilaceus -a -um cone-like, cone-shaped

strobilifer -era -erum cone-bearing

strobus an ancient name for an incense-bearing tree

strongyl-, strongylo- round-, rounded-

Strophanthus twisted flower (the elongate lobes of the corolla)

strumarius -a -um, strumosus -a -um cushion-like, swollen (refers to former medicinal use for treating swelling of the neck)

Struthiopteris ostrich feather (the fertile fronds)

stylosus -a -um with a prominent style

-stylus -a -um -styled

styracifluus -a -um flowing with gum

Suaeda from an Arabic name

suaveolens sweet-scented

suavis -is -e sweet, agreeable

sub-, suc-, suf-, sug- below-, under-, approaching-, nearly-, just less than-, usually-

subcaeruleus -a -um slightly blue

suberosus -a -um slightly bitten, corky

sublustris -is -e glimmering, almost shining

submersus -a -um submerged

subterraneus -a -um below ground, underground

subtilis -is -e fine

Subularia awl (the leaf-shape)

subulatus -a -um awl-shaped, subulate

succisus -a -um cut off from below (the rhizome)

succotrinus -a -um from Socotra, Indian Ocean

succulentus -a -um fleshy, soft, juicy

sucosus -a -um sappy

sudeticus -a -um from the Sudetenland of Czechoslovakia and Poland

suecicus -a -um from Sweden, Swedish

suffocatus -a -um suffocating (the flower heads turn to the ground)

suffructicosus -a -um somewhat shrubby at the base, soft-wooded and growing yearly from ground level

suionum of the Swedes

sulcatus -a -um furrowed, grooved

sulfureus -a -um, sulphureus -a -um pale yellow, sulphur-yellow

sultani for the Sultan of Zanzibar

sumatranus -a -um from Sumatra, Indonesia

superbus -a -um magnificent, superb

super-, supra- above-, over-

superciliaris -is -e eyebrow-like, with eyebrows

supinus -a -um lying flat, extended, supine

supranubius -a -um of very high mountains, from above the clouds

surculosus -a -um shooting, suckering

suspendus -a -um hanging down, suspended, pendant

sutchuensis -is -e from Szechwan, China

sy-, syl-, sym-, syn-, syr-, sys- with-, together with-, united-, joined-

sylvaticus -a -um, sylvester -tris -tre wild, of woods or forests, sylvan

sylvicola inhabiting woods

sympho-, symphy- growing together-

Symphoricarpus (os) clustered berries

Symphytum Dioscorides' name for healing plants

syn- united-, together-

syphiliticus -a -um (see *siphiliticus*)

Syringa pipe (use of the hollow stems for making flutes)

syzigachne with scissor-like glumes

Syzygium joined (from the form of branching and opposite leaves. Formerly applied to *Calyptranthes*)

Tabebuia from a Brazilian name

tabernaemontanus -a -um for J. T. Bergzabern

tabularis -is -e, tabuli- table-flat, flattened

tabuliformis -is -e flat and circular, plate-like

tacamahaccus -a -um from an Aztec name

tacazzeanus -a -um from the Takazze River, Ethiopia

taccifolius -a -um with leaves like *Tacca* (arrowroot)

taeda an ancient name for all resinous pines

taediger -era -erum torch-bearing

Tagetes for the grandson of Jupiter

taiwanensis -is -e from Formosa (Taiwan), Formosan

Tamarindus from the Arabic for Indian date

Tamarix the Latin name

tamnifolius -a -um byrony-leaved, with leaves like *Tamus*, Tamnus of Pliny

Tamus the name in Pliny for a kind of vine

Tanacetum immortality (tansy was placed amongst winding-sheets on the dead)

tanguticus -a -um of the Tangut tribe of N.E. Tibet

tapein-, tapeino- humble-, modest-

tapeti- carpel-like-

taraxaci- dandelion-like

Taraxacum from the Persian for a bitter herb

tardi-, *tardivus -a -um*, *tardus -a -um* slow-, late-

tartareus -a -um of the underworld (coloration)

tartaricus -a -um, tataricus -a -um from Tartary, central Asia

tauricus -a -um of the Crimea

taurinus -a -um from Turin, Italy, or of bulls

taxi- yew-like, resembling *Taxus-*

taxodioides resembling *Taxodium*

Taxodium yew-like

taxoides resembling yew

Taxus the Latin name for yew

tazettus -a -um small-cupped (the corona of *Narcissus*)

Tectona from the Tamil name for teak

tectorum of tiles, of rooftops, growing on rooftops

tectus -a -um with a thin covering, tectate

tef the Arabic name for *Eragrostis abyssinica* (tef grass)

tegetus -a -um mat-like

tel-, tele- far-, far off-

telephioides resembling *Sedum telephium*

telephium a Greek name for a plant thought to be capable of
   indicating reciprocated love, far off lover

Tellima an anagram of *Mitella*

telmataia, telmateius -a -um of marshes, of muddy water

telonensis -is -e from Toulon (Telenium), France

temenius -a -um of holy places

temulentus -a -um, temulus -a -um bewildered, intoxicated,
   drunken (toxic seed of ryegrass)

tenax gripping, stubborn, firm, persistent

tenebrosus -a -um delicate, somewhat tender

tenens enduring, persisting

teneri-, tener -era -erum soft, tender, delicate

tenuis-, tenuis -is -e slender, thin, fine

tenuior more slender

tephro- ash-grey

Tephrosia ashen (the leaf colour)

teres, teretis, terete quill-like, cylindrical

teretiusculus -a -um somewhat smoothly rounded

terminalis -is -e terminal (e.g. the flower on a stem)

ternateus -a -um of the Ternate Islands, Moluccas

ternatus -a -um with parts in threes, ternate (see Fig. 5(*e*))

terrestris -is -e growing on the ground, not epiphytic or aquatic

tessellatus -a -um chequered, mosaic-like

testaceus -a -um brownish-yellow

testicularis -is -e, testiculatus -a -um tubercled, testicle-shaped *tetra-* four-

Tetracme four points (the shape of the fruit)

Tetragonolobus quadrangular pod (the fruit)

tetragonus -a -um four-angled

tetrahit four times (tetraploid), foetid

tetralix a name used by Theophrastus for the crossleaved state when leaves are arranged in whorls of four

tetraplus -a -um fourfold (e.g. ranks of leaves)

Teucrium Dioscorides' name perhaps for Teucer, a hero and first King of Troy

texanus -a -um from Texas, U.S.A.

textilis -is -e used for weaving

thalassicus -a -um growing in the sea

thalianus -a -um after Dr Johannus Thalius

Thalictrum a name used by Dioscorides

-thamnus -a -um -shrub-like

Thapsus from the Island of Thapsos

thebaicus -a -um from Thebes

theco-, -thecus -a -um box-, -chambered

theifer -era -erum tea-bearing

thelo-, thely- female-, nipple-

Thelycrania the name used by Theophrastus

Thelypteris female fern

Theobroma food of god

theriacus -a -um of snakes (medicinal use in snakebite)

thermalis -is -e of warm springs

thero- summer-

Thesium a name in Pliny for a bulbous plant

thessalonicus -a -um, thessalus -a -um from Thessaly

thibeticus -a -um from Tibet

thirsi- panicled-

Thladiantha eunuch flower (aborted stamens in female flowers)

Thlaspi the name used by Dioscorides

thorus -a -um of corruption, of ruination

-thrinax -trident

-thrix -hair, -haired

Thuja Theophrastus' name for a fragrant-wooded tree

thurifer -era -erum incense-bearing

thuringiacus -a -um from mid-Germany

Thymus Theoprastus' name for a plant used in sacrifices

thyoides Thuja-like

thyrsoideus -a -um panicle-like, thyrsoid (see Fig. 3(d))

thysano- fringed-

tibeticus -a -um from Tibet
tibicinus -a -um flute-like
Tibouchina from a Guianan name
tigrinus -a -um striped, heavily spotted, tiger-toothed
Tilia the Latin name for the lime tree
tiliaceus -a -um lime-like, resembling *Tilia*
tinctorius -a -um used in dyeing, of dyeing
tinctus -a -um coloured
tingitanus -a -um from Tangiers, Morocco
tinus the old Latin name for laurustinus (*Viburnum*)
tipuliformis -is -e resembling a Tipulid, daddy-longlegs-like
tirolensis -is -e from the Tyrol, Tyrolean
titanius -a -um very large
titanum of the Titans, gigantic
Tithymalus the ancient name for plants with latex, spurges
tomentellus -a -um somewhat hairy
tomentosus -a -um thickly matted with hairs
tonsus -a -um shaven, sheared, shorn
Tordylium the name used by Dioscorides
Torilis an example of a meaningless name, by Adanson
torminalis -is -e of colic (used medicinally to relieve colic)
torminosus -a -um causing colic
torosus -a -um cyclindrical with regular constrictions
torridus -a -um of very hot places
tortilis -is -e, tortus -a -um twisted
tortuosus -a -um meandering (irregularly twisted stems)
torulosus -a -um swollen or thickened at intervals
torvus -a -um fierce, harsh, sharp
toxicarius -a -um, toxicus -a -um poisonous
toza from a South African native name
trachelium neck (use for throat infections)
trachelo- neck-
trachy- shaggy-, rough-
trachyodon short-toothed, rough-toothed
Trachystemon rough stamens
trago- goat-
Tragopogon goat beard
Tragus goat
trans- through-, across-, beyond-
transiens intermediate, passing over
translucens transparent
transwallianus -a -um from Pembroke, S. Wales
transylvanicus -a -um from Romania (Transylvania)

Trapa from *calitrapa*, a four-spiked club-like weapon
tremulus -a -um trembling, shaking
tri- three-
triandrus -a -um with three stamens
triangulari-, triangularis -is -e triangular (leaf shape)
tricho-, -thichus -a -um hairy-, hair-like-, -hairy
Trichomanes Theophrastus' name for maidenhair spleenwort
Trichophorum hair carrier (perianth bristles)
Trichosanthes hair flower (the fringed corolla lobes)
trichospermus -a -um hairy-seeded
tricolor three-coloured
tricornis -is -e, tricornutus -a -um three-horned, having three
    horn-like structures
tridactylites three-fingered
trientalis -is -e a third of a foot in length, about 4 inches tall
trifidus -a -um divided into three
Trifolium the name in Pliny for trefoil (trifoliate) plants
Triglochin three-barbed (the fruits)
Trigonella triangle (the flower of fenugreek seen from the front)
Trillium three-partite (the parts are conspicuously in threes.
    Lily-like)
trimestris -is -e of three months, maturing in three months
trionus -a -um three-coloured
Tripleurospermum three-ribbed seed
triplo- threefold-
triquetrus -a -um three-edged, three-angled (e.g. stems)
Trisetum three awns
tristis -is -e sad, dull-coloured
triternatus -a -um three times in threes (e.g. division of the
    leaves)
Triticum the classical name for wheat
tritifolius -a -um with polished leaves
trivialis -is -e common, ordinary, wayside, of the crossroads
trocho- wheel-, wheel-like-
troglodytarum of apes or monkeys (implies inferiority or unsuit-
    ability for man, cf. *sapientius*)
Trollius from the Swiss-German 'Trollblume' – rounded flower
Tropaeolum trophy (the gardener's *Nasturtium* was likened by
    Linnaeus to the shields and helmets displayed after victories in
    battle)
tropicus -a -um of the tropics
truncatulus -a -um, truncatus -a -um blunt-ended (e.g. leaf margin
    or apex, see Fig. (7(d))

Tsuga from the Japanese name for the Hemlock cedar
tubaeflorus -a -um with tubular flowers
tubatus -a -um trumpet-shaped
tuberculatus -a -um, *tuberculosus -a -um* warted, warty, tuber-
   culate (surface texture)
tubergenii for the van Tubergen bulb growers of Holland
tuberosus -a -um swollen, tuberous
tubifer -era -erum, *tubulosus -a -um* tubular, bearing tubular
   structures
tubiflorus -a -um trumpet-flowered
tucumaniensis -is -e from Argentina, Argentinian
tul- warted-
Tulipa from the Persian name for a turban
tulipi- tulip-like-
tumescens inflated
tumidi-, *tumidus -a -um* swollen, tumid
tunbrigensis -is -e from Tunbridge
tunicatus -a -um coated, having a tunic or covering
tuolumnensis -is -e from Tuolumne County, California, U.S.A.
turbinatus -a -um top-shaped, turbinate
turcicus -a -um from Turkey, Turkish
turcomanicus -a -um from Turkestan
turcumaniensis -is -e from Turku, Finland
turgidus -a -um inflated, turgid
turgiphalliformis -is -e erect-phallus-shaped
Turrita, *Turritis* tower
Tussilago coughwort (medicinal use of leaves for coughs)
tylo- knob-, callus-, wart-
Typha a Greek name for various plants
typhinus -a -um, *typhoides* bulrush-like, resembling *Typha*
typicus -a -um, *-topos*, *-typus* typical, the type

-ugo -having (a feminine suffix in generic names)
ulcerosus -a -um knotty, lumpy
-ulentus -a -um -abundant, -full
Ulex a name in Pliny
ulicinus -a -um resembling *Ulex*
uliginosus -a -um marshy, of swamps or marshes
-ullus -a -um -smaller, -lesser
Ulmaria elm-like (the appearance of the leaves)
ulmi-, *ulmoides* elm-like, resembling *Ulmus*
Ulmus the Latin name for elms
-ulus -a -um -tending to, -having somewhat

ulvaceus -a -um resembling the green seashore alga *Ulva*

umbellatus -a -um with the branches of the inflorescence all
   rising from the same point, umbelled (see Fig. 2(*e*))

umbilicus -a -um navelled

umbo- knob-like-

umbracul- umbrella-like-

umbrosus -a -um shade-loving, growing in shade, giving shade

uncus -a -um hooked, crooked

uncatus -a -um, uncinatus -a -um hook-tipped

undatus -a -um, undulatus -a -um wavy, not flat, undulate

unedo the Latin name for the *Arbutus* tree and its fruit meaning
   'I eat one'

ungui- clawed-, with a claw-

unguicularis -is -e, unguiculatus -a -um with a small stalk or claw
   (e.g. the petals)

uni- one-, with a single-

unilateralis -is -e one-sided, unilateral

unioloides resembling *Uniola* (American sea oats)

uplandicus -a -um from Uppland, Sweden

uragogus -a -um diuretic

urbanus -a -um, urbicus -a -um of towns

urceolatus -a -um pitcher- or urn-shaped

Urena from the Malabar name

urens stinging

Urera burner (cow itch)

Urginea from the name of an Algerian tribe

uro-, -urus -a -um tail-, -tailed

ursinus -a -um bear-like (the smell)

-usculus -a -um -ish (a diminutive ending)

Urtica sting (the Latin name)

usitatissimus -a -um most useful, everyday, ordinary

ustulatus -a -um scorched-looking

utilis -is -e useful

Utricularia bladderwort (the traps on the underwater stems)

utricularis -is -e, utriculatus -a -um with bladders or utricles,
   bladder-like

-utus -a -um -having

uvarius -a -um an old generic name, clustered, like grape
   bunches

uva-ursi berry of bears

uvifer -era -erum grape-bearing

vaccaria an old generic name, from 'vacca' a cow

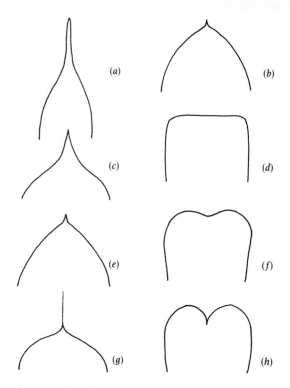

Fig. 7. Leaf-tip shapes which provide specific epithets:
(*a*) caudate (e.g. *Ornithogalum caudatum* Jacq.), with a tail;
(*b*) mucronate (e.g. *Erigeron mucronatus* DC.), with a hard tooth;
(*c*) acuminate (e.g. *Magnolia acuminata* L.), pointed abruptly;
(*d*) truncate (e.g. *Zygocactus truncatus* K.schum.), bluntly foreshort-
ened; (*e*) apiculate (e.g. *Braunsia apiculata* Schw.), with a short broad
point; (*f*) retuse (e.g. *Daphne retusa* Hemsl.), shallowly indented;
(*g*) aristate (e.g. *Berberis aristata* DC), with a hair-like tip, not always
restricted to describing the leaf apex; (*h*) emarginate (e.g. *Limonium
emarginatum* (Willd.) O.Kuntz), with a deep mid-line indentation.

vaccini- bilberry-like-

Vaccinium a name of great antiquity (like *Hyacinthus*) with no clear meaning

vaccinus -a -um of cows, the colour of a red cow

vacillans variable

vagans wandering, of wide distribution

vagensis -is -e from the River Wye

vaginatus -a -um having a sheath, sheathed

vagus -a -um uncertain, wandering

valdivianus -a -um from Valdivia, Chile

valentinus -a -um from Valencia, Spain

Valeriana health (from a medieval name for its medicinal use)

valesiacus -a -um from Valois, N.E. France

validus -a -um well-developed, strong

vallesiacus -a -um, vallesianus -a -um from Valais (Wallis), Switzerland

valverdensis -is -e from Valverde, Hierro, Canary Isles

vari-, varii-, varius -a -um differing, varying, changing

variabilis -is -e, varians variable

variatus -a -um various, several

variegatus -a -um irregularly coloured

variolatus -a -um pock-marked, pitted

vectensis -is -e from the Isle of Wight (Vectis)

vegetus -a -um strongly growing

velaris -is -e veiling

vellereus -a -um fleecy, densely long-haired

velox quick-growing

veluti-, velutinus -a -um with a soft silky covering, velvety

venator hunter (the flowers are 'hunting pink')

venenatus -a -um poisonous

venosus -a -um conspicuously veined

ventricosus -a -um bellied-out below, distended, expanded

venulosus -a -um finely veined

venustus -a -um graceful, beautiful, charming

Veratrum the Latin name

verbanensis -is -e from the area of Lake Maggiore

verbasci- Mullein-like, resemblng *Verbascum*

Verbascum a name in Pliny

Verbena the Latin name for the leafy twigs used in wreaths for ritual use and medicine

verbenacus -a -um, verbeni- vervain-like, resembling *Verbena*

verecundus -a -um modest

veris -is -e of spring (flowering time)

vermi-, vermicularis -is -e, vermiculatus -a -um worm-like

vernalis -is -a, vernus -a -um of spring (flowering time)

vernicifer -era -erum producing varnish

vernicosus -a -um varnished, glossy

vernix varnish

Veronica for St Veronica, who wiped the sweat from Christ's face

verrucosus -a -um warty, verrucose

versicolor varying or changeable in colour

verticillaris -is -e having whorls (several leaves or flowers all arising at the same level on the stem)

verticillaster with whorls of flowers

verticillatus -a -um arranged in whorls, verticillate

veruculosus -a -um somewhat warty

verus -a -um true, genuine

vescus -a -um small, feeble, fine, edible

vesicarius -a -um inflated, bladder-like

vesicatorius -a -um blistering

vesiculosus -a -um inflated, bladder-like

vespertilis -is -e bat-like (with two large lobes)

vespertinus -a -um of the evening (evening flowering)

vestalis -is -e white

vestitus -a -um covered, clothed (with hairs)

Vetiveria from a southern Indian vernacular name

Vetrix osier

vexans annoying, wounding

vexillaris -is -e with a standard (as the large 'sail' petal of a pea flower)

vialis -is -e, viarus -a -um of the wayside, ruderal

Viburnum the Latin name

Vicia the Latin name for a vetch

vicii- vetch-like-, resembling *Vicia*

vicinus -a -um neighbouring

victorialis -is -e victorious (protecting)

vilis -is -e common, of little value

villicaulis -is -e with a shaggy stem

villipes with a long-haired stalk

villosulus -a -um slightly hairy

villosus -a -um shaggy, with long soft hairs

vilmorinianus -a -um, vilmorinii for the French nurserymen, Vilmorin-Andrieux

viminalis -is -e, vimineus -a -um with long slender shoots, suitable for wicker or basketwork, osier-like *vinaceus -a -um* of the vine, wine-coloured

Vinca binding (the Latin name refers to its use in wreaths)

Vincetoxicum poison-beater (its supposed antidotal property to snakebite)

vincoides periwinkle-like, resembling *Vinca*

vinculans binding, fettering

vindobonensis -is -e from Vienna, Viennese

vinealis -is -e of vines and vineyards, growing in vineyards

vinicolor wine-red

vinifer -era -erum wine-bearing

vinosus -a -um wine-like, wine-red

Viola the Latin name for several fragrant-flowered plants

violaceus -a -um violet-coloured

violescens turning violet

virens green

virescens light green

virgatus -a -um twiggy, with straight slender twigs

Virgaurea rod of gold

virginalis -is -e, virgineus -a -um maidenly, virginal, purest white

virginianus -a -um from Virginia, U.S.A., Virginian

virginicus -a -um from the Virgin Islands, Virginian

virginiensis -is -e Virginian

virgultorum of thickets

viridescens turning green, becoming green

viridior more green, greener

viridi-, viridis -is -e youthful, fresh green

viridulus -a -um greenish

virmiculatus -a -um vermillion

virosus -a -um slimy, rank, poisonous

Viscaria bird lime (the sticky stems of German catchfly)

viscatus -a -um clammy

viscidi-, viscidus -a -um, viscosus -a -um sticky, clammy, viscid

Viscum the Latin name for birdlime and mistletoe

vitalba vine of white (the appearance of *Clematis* in the fruiting state)

vitellinus -a -um egg-yolk yellow

viti- vine-like-, resembling *Vitis*

viticellus -a -um like a small vine

vitiensis -is -e from the Fijian Islands (Viti Levu)

Vitis the Latin name for the grapevine

vitis -idaea vine of Mt Ida or Idaea, Greece

vitreus -a -um glassy

vittatus -a -um striped lengthwise

vittiformis -is -e band-like
vittiger -era -erum bearing lengthwise stripes
vivax long-lived (flowering for a long time)
viviparus -a -um producing plantlets (often in place of flowers or
  as precocious germination on the parent)
volgaricus -a -um from the Russian River Volga
volubilis -is -e entwining, enveloping
volutus -a -um with rolled leaves
vomitorius -a -um causing regurgitation, emetic
vulcanicus -a -um fiery, of extinct volcanoes
vulgaris -is -e, vulgatus -a -um common
vulnerarius -a -um of wounds (wound healing properties)
vulnerus -a -um of wounds
vulpinus -a -um of the fox, foxy (coloration but also implies
  inferiority)
vulvarius -a -um cleft, with two ridges

wardii for Frank Kingdon-Ward, collector of East Asian plants
warleyensis -is -e of Warley Place, Essex (home of Miss Ellen Ann
  Willmott)
watermaliensis -is -e from Watermal, Belgium
wolgaricus -a -um from the region of the River Volga

xalapensis -is -e from Xalapa, Mexico
xanth-, xanthii, xantho-, xanthinus -a -um yellow
Xanthium Discorides' name for cocklebur, from which a yellow
  hair-dye was made
xanthospilus -a -um yellow-spotted
xanthostephanus -a -um with a yellow crown
xanthoxylon yellow-wooded
xero- dry-
xerophilus -a -um drought-loving, living in dry places
xiphioides sword-like, shaped like a sword
Xiphium sword (the Greek name for a *Gladiolus*)
xiphochilus -a -um with a sword-shaped lip
xiphophyllus -a -um with sword-shaped leaves
xylo- wood-, woody-
xylocanthus -a -um woody-thorned

yakusimanus -a -um from Yakushioma, Japan
yedoensis -is -e from Tokyo (Yedo), Japan
yemensis -is -e from The Yemen, Arabia
yosemitensis -is -e of the Yosemite Valley, California, U.S.A.

Yucca from a Carib name formerly applied to Cassava
yuccifolius -a -um with *Yucca*-like leaves
yunnanensis -is -e from Yunnan, China

za- much-, many-, very-
zaleucus -a -um very white
zalil from an Afghan name for a *Delphinium*
Zamia a name in Pliny refers to the possession of cones
zamii- resembling *Zamia-*
Zanthoxylum yellow wood
zanzibarensis -is -e, zanzibaricus -a -um from Zanzibar, East
　Africa
zapota South American name for the chicle tree, *Sapodilla*
Zea from the Greek name of another cereal
zebrinus -a -um from the Portuguese, meaning striped with
　different colours
Zephyranthes west wind flower
zerphyrius -a -um western, flowering or fruiting during the
　monsoon season (for Indonesian plants)
Zerna a Greek name
zetlandicus -a -um from the Shetland Isles
zeylanicus -a -um from Ceylon, Singhalese
zibethinus -a -um of the civet (smelling of civet)
Zingiber from a pre-Greek name, possibly from India
zizanoides resembling *Zizania*, like Canadian wild rice
zonalis -is -e, zonatus -a -um girdled with distinct bands or
　concentric zones
zooctonus -a -um poisonous
zoster- girdle-
Zostera Theophrastus' name for a marine plant
zygo- yoked-, paired-, balanced-
Zygocactus jointed stem (the stems of the Christmas cactus
　appear to be articulated)
zygomeris -is -e with twinned parts
Zygophyllum yoked leaves (the leaves of some are conspicuously
　paired)

Bibliography

Adanson, M. *Familles des Plantes*. Paris 1763–1764.

Bauhin, C. (1623). *Pinax Theatri Botanici*. Basel.

Bailey, L. H. (1949). *Manual of Cultivated Plants*. Macmillan, New York.

Brickell, C. D. *et al.* (1980). International Code of Nomenclature for Cultivated Plants. In *Regnum Vegetabile*, **104**, Deventer.

Britten, J. & Holland, R. (1886) *A Dictionary of English Plant Names*. The English Dialect Society, London

Chittenden, F. J. (Ed.) (1951). *Royal Horticultural Society Dictionary of Gardening*. Vols. 1–4, and Supplements 1956 and 1969 (ed. P. M. Synge). Oxford University Press, Oxford.

Dioscorides, P. (1934). *Materia Medica*. John Goodyer translation of 1655, ed. R. T. Gunther. Oxford University Press, Oxford.

Farr, E. R. *et al.* (Eds.) (1979). Index Nominum Genericorum. In *Regnum Vegetabile*, **100**, **101** and **102**, The Hague.

Fernald, M. L. (1950). *Gray's Manual of Botany*. American Book Co., New York.

Gilbert-Carter, H. (1964). *Glossary of the British Flora*. 3rd edn. Cambridge University Press, Cambridge.

Green, M. L. (1927). The history of plant nomenclature. *Kew Bulletin*, **403–15**

Grigson, G. (1975). An Englishman's Flora. Hart-Davis, St Albans

Ivimey-Cook, R. B. (1974). *Succulents – A Glossary of Terms and Descriptions*. National Cactus and Succulent Society, Oxford.

Jackson, B. D. (1960). A Glossary of Botanical Terms. (4th edn). Duckworth, London.

Jeffrey, C. (1977). *Biological Nomenclature*. Edward Arnold, London.

Johnson, A. T. & Smith, H. A. (1958). *Plant Names Simplified*. Feltham

Jussieu, A. L. de (1789). *Genera Plantarum*. Paris.

Mentzel. C. (1682). *Index Nominum Plantarum Multilinguis (Universalis)*. Berlin.

Parkinson, J. (1629). *Paradisi in Sole*. Reprinted by Methuen, London (1904).

Plowden, C. C. (1970). *A Manual of Plant Names*. George Allen & Unwin, London.

Prior, R. C. A. (1879). *On the Popular Names of British Plants*. 3rd edn. London.

Rauh, W. (1979). *Bromeliads*. (English translation by P. Temple.) Blandford Press, Dorset

Schultes, R. E. & Pease, A. D. (1963). *Generic Names of Orchids – Their Origin and Meaning*. Academic Press, London.

Smith, A. W. (1972). A Gardener's Dictionary of Plant Names. (Revised and enlarged by W. T. Stearn.) Cassell, London.

Sprague, T. A. (1950). The evolution of botanical taxonomy from Theophrastus to Linnaeus. In *Lectures on the development of taxonomy*, Linnean Society of London.

Stafleu, F. A. *et al*. (Eds.) (1983). International Code of Botanical Nomenclature. In *Regnum Vegetabile*, **111**, Utrecht.

Stearn, W. T. (1983). *Botanical Latin*. David & Charles, Newton Abbot.

Willis, J. C. (1955). *A Dictionary of the Flowering Plants and Ferns*. (6th edn) Cambridge University Press, Cambridge.

Wilmott, A. J. (1950). Systematic botany from Linnaeus to Darwin. In *Lectures on the development of taxonomy*, Linnean Society of London.

Zimmer, G. F. (1949). *A Popular Dictionary of Botanical Names and Terms*. Routledge & Kegan-Paul, London.

DATE DUE

			PRINTED IN U.S.A.